N

MINERAL NAMES
WHAT DO THEY MEAN?

MINERAL NAMES
WHAT DO THEY MEAN?

by

Richard Scott Mitchell

assisted by

John Reese Henley

VNR VAN NOSTRAND REINHOLD COMPANY
NEW YORK   CINCINNATI   ATLANTA   DALLAS   SAN FRANCISCO
LONDON   TORONTO   MELBORNE

Van Nostrand Reinhold Company Regional Offices:
New York   Cincinnati   Atlanta   Dallas   San Francisco

Van Nostrand Reinhold Company International Offices:
London   Toronto   Melbourne

Library of Congress Catalog Card Number: 78-26141
ISBN: 0-442-24593-9

Manufactured in the United States of America

Published by Van Nostrand Reinhold Company
135 West 50th Street, New York, N.Y. 10020

Published simultaneously in Canada by Van Nostrand Reinhold Ltd.

15 14 13 12 11 10 9 8 7 6 5 4 3 2 1

Library of Congress Cataloging in Publication Data

Mitchell, Richard Scott, 1929-
    Mineral names.

    Bibliography: p.
    Includes index.
    I. Mineralogy—Names. I. Title.
QE357.M57   549′.01′4   78-26141
ISBN 0-442-24593-9

TO MY PARENTS

To my father, who encouraged my interest in minerals.
To my mother, who pointed to the One who formed them.

Foreword

Each science has a nomenclature of its own. Nearly all began with "trivial" names, i.e., names not produced by the use of a systematic procedure. Most sciences eventually dropped these "trivial" names and adopted names based on a system of nomenclature. Thus, for example, chemistry was forced nearly a century ago to discard names such as "sweet spirit of niter" and "Glauber's salt" and to adopt rules for the systematic naming of chemical compounds. The reason was simply that there were so many compounds (more than 3 million are now registered) that systematic rules for naming them were necessary. These systematic names have the enormous advantage that anyone who knows the rules can write the formula of the compound from its name.

In contrast, the relatively small number of necessary mineral names (probably less than 2500 even now, with about fifty new ones annually) has reduced the pressure for a systematic nomenclature in mineralogy. Furthermore, the conflicting views among mineralogists of the past century as to whether classification of minerals (and, therefore, their nomenclature) should be based on physical properties or chemical composition was not decided in favor of the latter until about 1850, by which time the "trivial" names were so firmly entrenched that they are still with us.

The activities of the Commission on New Minerals and Mineral Names of the International Mineralogical Association are directed mainly toward preventing the introduction of unnecessary or confusing new names, toward obtaining agreement on a single name when more than one is in use, and toward studying mineral groups with the aim of defining the names used and eliminating those that are unnecessary.

There is little prospect that a completely new systematic nomenclature of mineralogy will be adopted in the near future. We might as well make the best of this by enjoying the diversity of names now in use and by learning the chemical, physical, geographical, etymological, and other derivations of these names, which Professor Mitchell has summarized so well in this volume.

Michael Fleischer
Washington, D.C.

Preface

The purpose of this work is to bring together the derivations of all mineral names currently used in the science. Before now, these derivations were widely scattered throughout the literature, and a diligent search was required to find the meaning of the name in question. The last comprehensive book in English devoted to the subject of the derivation of mineral names was published over 80 years ago (in 1896, *A Dictionary of the Names of Minerals* by Albert H. Chester). Since that time hundreds of new minerals have been discovered and named and many of the older names have become obsolete.

The direct derivation of mineral names is primarily emphasized in this study, but it is not intended to be a study of word etymologies in the broader sense, where the history of a particular word, with all its changes of form, spellings, and meanings, is traced. We will see that most minerals have been named directly for persons, places, and things, as well as for chemical and physical properties.

In compiling this work, no attempt was made to evaluate the suitability of one name over another. In general, synonyms and obsolete names are omitted except where they add to the understanding of name origins or where they are used as varietal names. In addition to modern species names, commonly used names for mineral groups, series, and varieties are included, but rock names are not included.

The initial guide for selection of mineral names was *Glossary of Mineral Species* (1975) by Michael Fleischer. Several of the synonyms included in his book are omitted. Some mineral names were added, especially those for recently published new minerals and for common varietal names. The derivations of mineral names were obtained from many sources, the following being particularly useful: *Dana's System of Mineralogy* (5th, 6th, 7th editions; and Appendixes 1–3 to the 6th edition); *A Dictionary of the Names of Minerals* (1896) by Albert H. Chester; "New mineral names," by M. Fleischer in *American Mineralogist,* bimonthly; "Revue des espèces minérales nouvelles," by C. Guillemin, F. Permingeat, and R. Pierrot in *Bulletin de la société française de minéralogie et de cristallographie,* quarterly; "Annual lists of new mineral names," by L. J. Spencer, M. H. Hey, *et al.* in *Mineralogical Magazine;* "New minerals," in *Mineralogical Abstracts,* quarterly.

In some cases where it was impossible to determine the origin of the name from the published data, letters were written to persons who might be in a position to know the derivations. Although the response to these requests was usually excellent, this method failed to obtain information on the few mineral names given in Appendix I.

Unfortunately it proved impossible to give equal treatment to biographical data for each person whose name is honored by a mineral term. Information is not available for many of the persons honored in this manner, especially for some of the very early names as well as for more recent persons who are not otherwise well-known. Although many sources of information were consulted for biographical data, special help was obtained from *World Who's Who in Science* (1968), edited by Allen G. Debus; *The National Union Catalog* of the Library of Congress and American Library Association; and from annual memorials published by the Mineralogical Society of America and the Geological Society of America.

The fact that many mineral names have been derived from the names of diverse localities over a period of several centuries leads to difficulties in achieving uniformity in any compilation. Locality names and spellings differ in time or with boundaries, and with language. Therefore no attempt was made to revise and update all locality data to conform with present political designations. For example, the names *Soviet Union* and *Leningrad* seem awkward to use when one is discussing mineralogists and minerals from the nineteenth century. In instances of this kind it was deemed more appropriate to retain such old names as St. Petersburg instead of the modern Leningrad. As further instances, many minerals were named when Zaire was called the Belgian Congo; this old term and similar ones are retained in numerous entries. Such Central and Eastern Europe regional names as Bavaria, Bohemia, Saxony, Tyrol, and many others, are generally used without regard to the present country to which they belong. As a rule, the locality designations given in the original published papers are used in this work.

Although mineral name derivations are not given in his work, Hugo Strunz in his *Mineralogische Tabellen* (5th edition) (1970) indicates when and by whom each mineral was named. Because this information has been covered adequately by him, it is not incorporated in the present study.

Scientific mineralogical data are adequately treated in many recent excellent books, and are not repeated here. Such data can be found in the following recent books: *Encyclopedia of Minerals* (1974) by W. L. Roberts, G. R. Rapp, Jr., and J. Weber; *Color Encyclopedia of Gemstones* (1977) by Joel E. Arem; *Glossary of Mineral Species* (1975) by Michael Fleischer; *Mineralogische Tabellen* (5th edition) (1970) by Hugo Strunz; *Crystal Chemical Classification of Minerals* (1972) by A. S. Povarennykh.

I wish to acknowledge the valuable aid of my friend John Reese Henley in various aspects of this compilation. His assistance was especially helpful in determining the derivations of the more common mineral names in the early stages of this work, in sorting out and classifying the various methods of naming minerals, and in compiling the index. I gratefully acknowledge the comments and corrections made by Michael Fleischer, who read an early draft of the manuscript. I also wish to express my appreciation to the numerous persons who responded to inquiries concerning name derivations not found in the literature.

The author will welcome corrections and additions, in the hope of making future editions of this work more accurate and comprehensive.

RICHARD S. MITCHELL
University of Virginia
Charlottesville

Contents

PART II: MINERAL NAMES: AN ALPHABETICAL LIST

MINERAL NAMES
WHAT DO THEY MEAN?

Part I
Mineral Names: A Discussion

1

Introduction

Names have been given to minerals since antiquity. Originally, names were derived from geographical localities or from mineral properties, either real or imaginary. Many of these ancient names have long etymologies and their earliest origins are obscure. With the advancement of the science of mineralogy, especially after Georgius Agricola (1494–1555), an ever increasing number of names appeared in mineralogical literature. According to Mandarino [1] less than 100 valid mineral species were known prior to 1800, although many more names were used as synonyms. A study of minerals named each year from 1800 to the present shows that the number introduced annually generally increases. Mandarino [1] summarized the number of new species introduced during every twenty-year period from 1800. From 1800 through 1819 the number was 87. In the period from 1820 through 1919 every twenty-year interval recorded on the average about 185 new minerals. In recent years the number has greatly increased. From 1920 through 1939 there were 256 new minerals described; from 1940 through 1959 there were 342; and in the thirteen-year period from 1960 through 1973 there were 575.

Along with the introduction of new minerals, old mineral names are constantly being reexamined for validity and many have been discredited. Names are usually discarded when it is shown they apply to minerals which are either identical to or are chemical variants of well-established species. These names then become obsolete, and are not to be used again. Of the more than 15,000

3

mineral names in the older literature,[2] only around 2,000 are for valid species while some additional ones have been retained for use as varietal names (for example, the varieties of quartz known as agate, amethyst, carnelian, chalcedony, and citrine). The total number of mineral species is relatively small compared to the number of known and theoretically possible synthetic inorganic substances. Povarennykh [3] indicates that by the year 1971 there were 1916 more or less firmly established mineral species. Furthermore, at that date there were 148 minerals which had not been examined adequately or were doubtful regarding their status as species, although probably about half of these would prove to be genuine species with further study.

Unfortunately, an old science like mineralogy is dominated by tradition and custom, and these factors have had considerable influence in retarding a more discerning approach to mineral nomenclature. There are two contrasting tendencies in mineralogical nomenclature, the rational and the irrational.

Rational mineral names convey information about the mineral itself. Although in the minority, many such names exist, as will become clear later. Names based on chemical compositions, crystal forms and habits, colors, lusters, and other properties, all illustrate rational nomenclature. On the other hand irrational names communicate very little or nothing about the mineral. The influence of Abraham G. Werner (1750–1817) at the end of the eighteenth century contributed to the predominance of the irrational principle when he began to name minerals for various persons. His method established a historical perspective in mineral names, but the names bore little or no relationship to the minerals themselves. In some instances the name indicates the discoverer or first analyst of the mineral. Names based on geographical localities convey slightly more information about the mineral, depending upon the kind of locality name used. The name franklinite, for Franklin, New Jersey, is considerably more specific than a name like brazilianite, for the entire country of Brazil. Over 1500 of the approximately 2600 names discussed in this study are based on the irrational principle of naming.

In 1959 the International Mineralogical Association established a

Commission on New Minerals and Mineral Names. Most journals now require that all new minerals and mineral names be approved by this commission before they are accepted for publication.[4,5] As a rule, manuscripts proposing new names for imperfectly or incompletely described minerals, or new names for compositional varieties, are not accepted. Some principles used in formulating new mineral names are discussed in Chapter 12. The commission also constantly reviews broader aspects of mineralogical nomenclature, including mineral synonyms and other mineralogical terminology.

REFERENCES

1. Mandarino, J. A. "Old mineralogical techniques." *Canadian Mineralogist* **15:** 1–2 (1977).
2. Aballain, M., Chambolle, P., Derec-Poussier, F., Guillemin, C., Mignon, R., Pierrot, R., and Sarcia, J. A. *Index Alphabétique de Nomenclature Minéralogique.* Paris: Bureau de Recherches Géologiques et Minières, 1968.
3. Povarennykh, A. S. *Crystal Chemical Classification of Minerals,* Volumes I–II. New York: Plenum Press, 1972.
4. Fleischer, Michael. "Procedure of the International Mineralogical Association Commission on New Minerals and Mineral Names." *American Mineralogist* **55:** 1016–1017 (1970).
5. Donnay, Gabrielle, and Fleischer, Michael. "Suggested outline for new mineral descriptions." *American Mineralogist* **55:** 1017–1019 (1970).

2
Names from Persons

A study of modern mineral species names will show that over 1100 of these were derived from personal names, usually family names, although in some instances first or middle names have been used. Apparently Abraham G. Werner (1750–1817) was the first to introduce personal names into mineralogy. In a discussion of mineralogical nomenclature, Dana [1] suggests that the earliest example was probably prehnite (after Col. Hendrik von Prehn), first applied by Werner in 1783. At about the same time he also named torbernite (after Torbern O. Bergman) and witherite (after William Withering). At first Werner's innovation was not well received. The French chemist and mineralogist Balthazar Georges Sage (1740–1824) protested [2] against the name prehnite (in 1789) and the use of personal names in general. In 1790 Franz Joseph Anton Estner (1739–1803), a mineralogist in Vienna, issued a pamphlet [3] against Werner in which he made light of his labors in the science, and under the heading of prehnite he ridiculed this method of naming, which he accused of creating a paternity, and providing the childless with children to hand down their names to posterity. But Dana [1] observed that "such names were, however, too easily made, too pleasant, as a general thing, to give and receive, and withal too free from real objection, to be thus stopped off, and they have since become numerous, even Vienna contributing her full share toward their multiplication." One might further note that minerals were never named for the critics Sage or Estner.

There are many modern critics of the use of personal names for minerals. One of these is Povarennykh,[4] who is especially concerned with the fact that these names have no relationship to the properties of the minerals. He classifies this naming system as irrational. He too is critical of the influence of Werner, and suggests that such mineral names were used to flatter members of the nobility and other influential people. This practice, according to Povarennykh,[4] could be understood when the financing of early science was almost entirely dependent on the generosity of aristocrats, as in Germany or Russia. Povarennykh would like to see a rational system developed for naming minerals, and suggests principles for a scientific nomenclature based on crystal-chemical properties.

Although it is true that some from the aristocracy are included in mineral names, most names honor scientists and other professionals. Among earth scientists, minerals have been named for mineralogists, crystallographers, geologists, petrologists, geochemists, mining engineers, mineral assayers, curators, collectors, and dealers. Other scientists include chemists, physicists, mathematicians, astronomers, metallurgists, botanists, zoologists, entomologists, physicians, and others. Also included are writers, poets, editors, philosophers, civil servants, explorers, aviators, missionaries, ministers, and various rulers and political leaders.

The national origins of the numerous persons who have been honored by mineral names encircle the Earth. In the Americas, persons from the United States, Canada, Mexico, Argentina, Brazil, Guyana, Chile, and Peru have been honored. This honor has also been given to persons from Australia, Japan, China, and India, as well as others from the various African nations. Numerous men from many of the European countries are included: England, Scotland, Ireland, Sweden, Norway, Denmark, Finland, Germany, France, Italy, Spain, Portugal, Belgium, Austria, the Soviet Union, Poland, Hungary, Czechoslovakia, Bulgaria, Greece, and others. Because the number of persons honored by having minerals named for them is so very large, the following discussion will include only a relatively small number of illustrative examples.

The purpose of this section is to illustrate the use of the names of distinguished mineralogists and other scientists in mineral nomenclature. Contributions of such persons to the development of the science will also be summarized. This is not meant to be a history of mineralogy, since our presentation will be limited to those individuals honored by mineral names. In order to sketch a more complete historical picture, however, a few obsolete mineral terms have been included here. Although their histories are entwined, we will concentrate first on European investigators and then on American investigators. All of the scientists to be considered were born before the twentieth century.

EUROPEAN MINERALOGISTS AND OTHER SCIENTISTS

The following mineral names are derived from the names of European scientists, many of whom are claimed by several scientific disciplines including mineralogy, geology, chemistry, and physics. The selected terms are arranged according to the birth date of the person to be honored.

agricolite (obsolete = eulytite), Georgius Agricola [Georg Bauer] (1494–1555), German mineralogist, often called the "father of mineralogy"; his *De Natura Fossilium* (1545) is the first handbook of mineralogy in the modern sense of the term; also authored the famous *De Re Metallica* (1556).

stenonite, Nicolaus Steno [Niels Steensen] (1638–1686), Danish anatomist and geologist; pioneered in geology, paleontology, and crystallography; discovered the law of the constancy of interfacial angles on crystals; published numerous books, including *De solido intra solidum naturaliter contento dissertationis prodromus* (1669).

linnaeite, Carolus Linnaeus [Carl von Linné] (1707–1778), Swedish botanist and taxonomist; often called the "father of modern systematic botany"; also contributed to the classification of minerals in his *Systema Naturae* (1735), and made other geological contributions.

torbernite, Torbern Olof Bergman (1735–1784), Swedish chemist and mineralogist; laid the foundations of mineralogical chemistry by classifying minerals according to chemical characteristics rather than appearances alone; studied crystallography; made quantitative determinations of mineral compositions.

romeite, Jean Baptiste Romé Delisle (1736–1770), French crystallographer; measured crystal angles; made over 500 drawings of crystals; published many works on crystallography, including *Essai de cristallographie* (1772); helped found the science of crystallography.

hauyne, René Just Haüy (1743–1822), French crystallographer and mineralogist; generally considered the founder of crystallography; published *Traité de Cristallographie* (1822, 3 vols.) and *Traité de Minéralogie* (1801, 5 vols.), and other works.

gahnite, Johan Gottlieb Gahn (1745–1818), Swedish chemist and mineralogist; student of Torbern Bergman; mine owner and mineral analyst; trained J. J. Berzelius in the use of the blowpipe for qualitative analyses of minerals.

wernerite (synonym for scapolite), Abraham Gottlob Werner (1750–1817), German mineralogist and geologist; established the separation of geology from mineralogy; published several books, including *Von den ausserlichen Kennzeichen der Fossilien* (1774). Mineral names, based on personal names, were his invention.

smithsonite, James Smithson (1765–1829), British chemist and mineralogist; studied the chemistry of minerals, for example, minium and zeolites; never visited the United States, but left his estate to a nephew, providing that should the nephew die childless, the estate would go to the United States to found the Smithsonian Institution (founded 1846).

wollastonite, William Hyde Wollaston (1766–1828), English chemist and physicist; his many contributions to science included the reflecting goniometer to measure crystal angles; placed crystallography on a firm quantitative base; established the Wollaston medal for research in mineralogy.

brookite, Henry James Brooke (1771–1857), English crystallographer and mineralogist. Brooke described many new minerals.

thomsonite, Thomas Thomson (1773–1852), Scottish chemist. He published many works on chemistry, including *Outlines of Mineralogy* (1836); introduced many new mineral species to the literature.

jamesonite, Robert Jameson (1774–1854), Scottish mineralogist and geologist; studied under A. G. Werner; published several books including *System of Mineralogy* (1804–1808, 3 vols.).

biotite, Jean Baptiste Biot (1774–1862), French physicist; studied magnetism and optics; with David Brewster, discovered optical biaxiality in crystals.

cordierite, Pierre Louis Cordier (1777–1861), French geologist and mining engineer; student of Haüy; helped enlarge the geological collection of Museum of Natural History in Paris.

davyne, Humphry Davy (1778–1829), English chemist; famous for his many discoveries in chemistry, and of the effects of laughing gas; isolated numerous metals for the first time; demonstrated that diamond is carbon.

gaylussite, Joseph Louis Gay-Lussac (1778–1850), French chemist and physicist; made many discoveries, including the law of combining volumes (Gay-Lussac's law), and the law which states that all gases expand equally for equal increments of temperature; analyzed composition of water; prepared many compounds and isolated elements; published numerous papers, some with Justus von Liebig and Alexander von Humboldt.

berzelianite, berzeliite, Jöns Jacob Berzelius (1779–1848), Swedish chemist; analyzed some 2000 compounds, including many minerals, validating the law of definite proportions; studied chemical radicals; discovered and isolated several elements; advocated the use of a chemical system of mineral classification; developed many analytical methods.

hausmannite, Johann F. L. Hausmann (1782–1859), German mineralogist; devised a system of crystallography based on spherical trigonometry; published many books including *Handbuch der Mineralogie* (1813, 3 vols.).

breithauptite, Johann August F. Breithaupt (1791–1873), German mineralogist; described pseudomorphism; discovered over 40 mineral species; made over 3000 measurements of calcite crystal faces; author of several books on mineralogy, including *Vollständiges Handbuch der Mineralogie* (1836–1847, 3 vols.).

dufrenite, dufrenoysite, Pierre Armand P. Dufrenoy (1792–1857), French mineralogist and geologist; introduced a new system of mineral classification based on crystallography; published several books including *Treatise on Mineralogy* (1847, 4 vols.)

whewellite, William Whewell (1794–1866), English natural scientist and mineralogist; wrote essays on mineralogical classification and nomenclature; made numerous contributions in geology (tides), mathematics, and history of sciences.

haidingerite, Wilhelm Karl von Haidinger (1795–1871), Austrian mineralogist and geologist; published many books on mineralogy including *Handbuch der bestimmenden Mineralogie* (1845); pioneered in crystallography, especially mineral optics; studied meteorites; described numerous new minerals.

adamite, Gilbert Joseph Adam (1795–1881), French mineralogist; described many new minerals; author of *Tableau Minéralogique* (1869); assembled a remarkable mineral collection.

delafossite, Gabriel Delafosse (1796–1878), French crystallographer; as a student of R. J. Haüy, he too was a founder of crystallography; wrote several books, including *Traité de cristallographie* (1821) and *Traité des minéralogie* (1822).

roselite, Gustav Rose (1798–1873), German mineralogist; a student of J. J. Berzelius; published several books on mineralogy, crystallography, meteorites, and on his travels; his book, *Das Krystalochemische Mineralsystem* (1852) was the first modern adoption of the system suggested by Berzelius; described and classified rocks.

wöhlerite, Friedrich Wöhler (1800–1882), German chemist; a one-time laboratory assistant to J. J. Berzelius; a pioneer in the study of organic chemistry, he synthesized urea; the first to isolate several metals (aluminum, beryllium); many publications including *Mineral Analyse in Beispielen* (1861); was associated with Justus von Liebig.

millerite, William Hallowes Miller (1801–1880), British mineralogist; authored works on crystallography, hydrostatics, and hydrodynamics; developed the Miller index for defining crystal faces.

liebigite, Justus von Liebig (1803–1873), German chemist; pioneered in organic chemistry; founded (Liebig's) *Annalen der Chemie;* introduced first student chemistry laboratories in continental Europe.

kobellite, Wolfgang Franz von Kobell (1803–1882), German mineralogist; noted for contributions to blowpipe methods; published book on mineral names and nomenclature (*Die Mineral-Namen und die Mineralogische Nomenklatur,* 1853).

bunsenite, Robert Wilhelm Bunsen (1811–1899), German chemist; with Kirchhoff, developed spectrum analysis (1859); outlined principles of chemical analysis by spectral methods; discovered cesium and rubidium; investigated geysers in Iceland; invented the gas (Bunsen) burner (1855).

rammelsbergite, Karl Friedrich Rammelsberg (1813–1899), German mineralogist and chemist; produced research and publications in the area of inorganic and mineralogical chemistry; author of numerous books, including *Handwörterbuch des Chemischen Teils der Mineralogie* (1841) and *Handbuch der Mineralchemie* (1860).

descloizite, Alfred Louis Olivier Legrand Des Cloizeaux (1817–1897), French mineralogist; did research on pseudomorphism and on optical properties of crystals; examined optical properties of over 450 minerals and salts; numerous publications, including *Manuel de Minéralogie* (1862, 2 vols.) and over 100 memoirs on mineralogy.

nordenskiöldine, Nils Adolf Erik Nordenskiöld (1832–1901), Finnish-Swedish geologist and explorer; a leader in Arctic exploration; first to sail through Northeast Passage; studied history of cartography; several books.

tschermakite, Gustav Tschermak von Seysenegg (1836–1927), Austrian mineralogist; did research on minerals (especially mica and feldspar groups), meteorites, crystallography; proved the plagioclase feldspars are an isomorphous series.

röntgenite, Wilhelm Konrad Röntgen (1845–1923), German physicist; best known for discovery and study of X-rays; conducted research in pyroelectricity, piezoelectricity, optical and electrical properties of quartz.

becquerelite, Antoine Henri Becquerel (1852–1908), French physicist; discovered radioactivity in uranium and its salts; investigated magnetism, phosphorescence, light polarization and absorption in crystals.

schoenfliesite, Artur M. Schoenflies (1853–1928), German mathematician; derived the 230 space groups in crystallography; published *Kristallsysteme und Kristallstruktur* (1891).

curite, Pierre Curie (1859–1906), French physicist and chemist; did research in crystallography; collaborated with his wife on the study of radioactivity; also on the discovery of polonium and radium.

braggite, both William Henry Bragg (1862–1942) and William Lawrence Bragg (1890–1971), British physicists and crystallographers; among many contributions, pioneered the study of X-ray crystallography; first to determine the arrangement of atoms in crystals; numerous books, including *X-rays and Crystal Structure* (1915), and other publications.

lacroixite, François A. A. Lacroix (1863–1948), French mineralogist; published *Minéralogie de la France et des ses colonies* (1893–1913, 5 vols.) and *Minéralogie de Madagascar* (1922–1923, 3 vols.); research on volcanic eruptions and meteorites; discovered over 40 new minerals.

spencerite, Leonard James Spencer (1870–1959), English mineralogist; wrote papers and books, including a translation of Max Bauer's *Edelsteinkunde* (*Precious Stones*) (1904); editor of *Mineralogical Magazine* (from 1900); keeper of minerals at British Museum of Natural History.

rutherfordine, Ernest Rutherford (1871–1937), British physicist; made numerous contributions to atomic physics; studied radioactivity of natural and artificial substances; calculated ages of uranium-bearing minerals; numerous books on radioactivity.

laueite, Max von Laue (1879–1960), German physicist; first used crystals as diffraction gratings for X-rays, proving electromagnetic nature of X-rays and providing technique for the study of the structures of crystals; also conducted research in many other areas.

AMERICAN MINERALOGISTS

Although mineralogy as a science originated in Europe, it achieved considerable growth in the United States and Canada throughout the nineteenth century. Many of the outstanding contributors to the development of the science in America had minerals named for them. Some of these, listed according to the birth dates of the men honored, are given here:

gibbsite, George Gibbs (1776–1833), American geologist; first vice-president of the American Geological Society (founded 1819); owner of an outstanding mineral collection acquired by Yale early in the nineteenth century.

troostite (variety of willemite), Gerard Troost (1776–1850), Dutch-American chemist and geologist; first president of Philadelphia Academy of Natural Sciences; professor of geology and mineralogy at University of Nashville (1828–1850) and Tennessee state geologist (1831–1850).

brucite, Archibald Bruce (1777–1818), American physician and mineralogist; assembled a valuable mineralogical cabinet in Europe during a tour; secured the charter for the College of Physicians and Surgeons of the State of New York (1807); founded the *American Mineralogical Journal* (1810), the first journal in America devoted to geological science (lasted four years); discovered and described the mineral which now bears his name.

sillimanite, Benjamin Silliman (1779–1864), American chemist and geologist; professor of chemistry and natural history at Yale College (1802–1853); gave first lectures in mineralogy and geology (1813); founder and first editor of *American Journal of Science* (1818).

cleavelandite (variety of albite), Parker Cleaveland (1780–1858), American geologist; professor of natural philosophy, chemistry, and mineralogy at Bowdoin College (1805–1858); wrote *Elementary Treatise on Mineralogy and Geology* (1816), the first American book on the subject.

vanuxemite (obsolete = hemimorphite), Lardner Vanuxem (1792–1848), American geologist; professor of chemistry and mineralogy at Columbia (S.C.) College (1819–1827); made geological surveys of New York, Ohio, Kentucky, Tennessee, and Virginia (1827–1830); a founder of the Association of American Geologists and Naturalists (1840), which became the American Association for the Advancement of Science.

danaite (obsolete = cobaltoan arsenopyrite), James Freeman Dana (1793–1827), American geologist. With Samuel L. Dana he published *Outlines of the Mineralogy and Geology of Boston and Vicinity* (1818).

mitchellite (obsolete = chromian spinel), Elisha Mitchell (1793–1857), American geologist, botanist, and naturalist; professor at the University of North Carolina (1818–1857); author of *Elements of Geology* (1842).

hitchcockite (obsolete = plumbogummite), Edward Hitchcock (1793–1864), American geologist and clergyman; professor at Amherst College (Massachusetts) (1825–1864); made first complete geological survey of

Massachusetts (1830); state geologist of Vermont (1857–1861); author of several books on geology.

weloganite, William Edmond Logan (1798–1875), Canadian geologist; a director of the Canadian Geological Survey (1842–1870); with T. S. Hunt, wrote *Geology of Canada* (1863).

rogersite (obsolete = churchite), William Barton Rogers (1804–1882), American geologist and educator; professor of natural history, University of Virginia (1835–1853); first Virginia state geologist (1835–1842); a founder (1861) and first president of Massachusetts Institute of Technology; author of many geological publications.

danalite, James Dwight Dana (1813–1895), American mineralogist and geologist; assisted Benjamin Silliman at Yale (1836–1837); participated in the Wilkes Exploring Expedition (1838–1842); editor of *American Journal of Science* (from 1846); professor of natural history and geology at Yale (1850–1890); author of many books including *System of Mineralogy* (first edition in 1837 at age 24).

genthelvite (from Genth + helvite), Frederick Augustus Genth (1820–1893), German-American mineralogist and chemist; assisted R. W. Bunsen in Marburg, Germany; opened chemical laboratory in Philadelphia (1848); professor of chemistry, University of Pennsylvania (1872–1888); described numerous new mineral species; wrote first mineralogy of Pennsylvania.

dawsonite, John William Dawson (1820–1899), Canadian geologist; principal of McGill University (1855–1893); wrote numerous books, especially on paleobotany.

cookeite, Josiah Parsons Cooke (1827–1894), American chemist and educator; professor of chemistry and mineralogy at Harvard (1850–1894); made first attempt to classify elements according to atomic weight; published numerous books on chemistry.

brushite, George Jarvis Brush (1831–1912), American mineralogist; assisted in chemistry at University of Virginia (1852–1853); professor of metallurgy and then mineralogy at Yale (1855–1912); wrote *Manual of Determinitive Mineralogy* (1874); assembled an extensive mineral collection and private library.

powellite, John Wesley Powell (1834–1902), American ethnologist, geologist, and administrator; professor at Illinois Wesleyan College (1865–1867) and at Illinois Normal University (1867); director of U.S.

Geological Survey and U.S. Bureau of Ethnology; explored Grand Canyon (1869) and Rocky Mountains (1871–1879).

wardite, Henry Augustus Ward (1834–1906), American naturalist and collector; professor of natural sciences, Rochester University (1860–1865); formed large cabinets of mineralogy and geology to be distributed to universities throughout the United States (1870–1900); founded Ward's Natural Science Establishment.

clarkeite, Frank Wigglesworth Clarke (1847–1931), American chemist; after teaching at Howard University and the University of Cincinnati, he became chief chemist at the U.S. Geological Survey and honorary curator of minerals, U.S. National Museum in 1883; made important contributions to geochemistry; author of numerous books.

hiddenite (variety of spodumene), William Earl Hidden (1853–1918), American mineralogist; explored and mined deposits of rare minerals, especially those of commercial value; investigated minerals of Alexander County, North Carolina, and Llano County, Texas; published articles on minerals.

penfieldite, Samuel Lewis Penfield (1856–1905), American mineral chemist and mineralogist; assisted George Brush at Yale; in 1881 took a full-time position there, becoming professor in 1893; described numerous mineral species and published over 80 papers on mineralogy and crystallography.

kunzite (variety of spodumene), George Frederick Kunz (1856–1932), American gemologist and mineralogist; associated with Tiffany and Company (from 1879); also worked with the U.S. Geological Survey and the American Museum of Natural History (N.Y.); took part in many expeditions and was author of numerous publications on gems, including *Gems and Precious Stones of North America* (1890).

mosesite, Alfred J. Moses (1859–1920), American mineralogist; professor at Columbia University (N.Y.); wrote the first American treatise on physical crystallography (*The Characters of Crystals,* 1899).

englishite, George L. English (1864–1944), American collector and mineral dealer; discovered several new minerals later described by others; author of several publications, including *Getting Acquainted with Minerals;* led many expeditions to collect specimens.

A vast number of twentieth century and contemporary mineralogists throughout the world have been honored by having minerals named for them, as shown by the names and dates of the elected presidents of the Mineralogical Society of America since its founding in 1919.

krausite, Edward H. Kraus, 1920.

palacheite (obsolete = botryogen), Charles Palache, 1921.

wherryite, Edgar T. Wherry, 1923.

eakleite (obsolete = xonotlite), Arthur S. Eakle, 1925.

schallerite, Waldemar T. Schaller, 1926.

austinite, Austin F. Rogers, 1927.

esperite, larsenite, Esper S. Larsen, Jr., 1928.

parsonsite, Arthur L. Parsons, 1929.

merwinite, Herbert E. Merwin, 1930.

whitlockite, Herbert P. Whitlock, 1933.

rossite, Clarence S. Ross, 1935.

bayleyite, William S. Bayley, 1936.

shortite, Maxwell N. Short, 1939.

foshagite, foshallasite, William F. Foshag, 1940.

buddingtonite, Arthur F. Buddington, 1942.

schairerite, John F. Schairer, 1943.

landesite, Kenneth K. Landes, 1945.

buergerite, Martin J. Buerger, 1947.

pavonite (from Latin *pavo,* peacock), Martin A. Peacock, 1948.

tunellite, George Tunell, 1950.

pabstite, Adolf Pabst, 1951.

fleischerite, Michael Fleischer, 1952.

hendricksite, Sterling B. Hendricks, 1954.

cliffordite, frondelite, Clifford Frondel, 1956.

djerfisherite, D. Jerome Fisher, 1957.

murdochite, Joseph Murdoch, 1960.

hurlbutite, Cornelius S. Hurlbut, 1963.

berryite, Leonard G. Berry, 1964.

faustite, George T. Faust, 1965.

brianite, stenhuggarite (Swedish *stenhuggar* = *stonemason*), Brian H. Mason, 1966.

jimthompsonite, James B. Thompson, Jr., 1968.

julgoldite, Julian R. Goldsmith, 1971.

yoderite, Hatten S. Yoder, Jr., 1972.

joesmithite, Joseph V. Smith, 1973.

Most of the presidents not included in the above list have names which might have caused some conflicts with minerals already named, for example: Thomas L. Walker, 1922 (there is an obsolete walkerite named for John Walker); Henry S. Washington, 1924 (obsolete washingtonite, named for Washington, Connecticut); Alexander H. Phillips, 1931 (phillipsite, named for William Phillips).

OBSOLETE NAMES DERIVED FROM MINERALOGISTS' NAMES

Many minerals named for very important or famous figures in the history of mineralogy are now discredited as species, and the names themselves have become obsolete. Such names were applied to specimens which were thought to be unique; however, with the advent of more sophisticated methods of analysis it was shown that these so-called species were simply chemical variants of other minerals or mixtures of minerals.

When a mineral name is discredited it cannot be used again to describe a different mineral. However, in recent years there has been a trend to circumvent an obsolete earlier name, in order to honor an identical or similar family name, by modifying the name in some way. For example the old name williamsite (variety of antigorite), named for L. White Williams in 1848, was bypassed in 1925 by using the name afwillite in honor of Alpheus F. Williams. In a recent paper, Fairbanks [5] expressed the hope that distinguished persons whose names may no longer be used as mineral names may again be honored by the use of composite names. Although Fairbanks does not object to names derived from amateur mineralogists and collectors, he deems it inappropriate to do so when so many discredited names of famous mineralogists are not being given a second chance.

Fairbanks' criticism should be considered. Not only should some of the names which have become obsolete be given another chance—for example, Theophrastus (ca. 372–ca. 287 B.C.), Georgius Agricola (1494–1555), Friedrich Mohs (1773–1839), C. U. Shepard (1804–1886), and Charles Palache (1869–1954), but names of other important contributors to the early history of mineralogy which have never been used, should be considered. Some of these might include Albertus Magnus (1193–1280), Roger Bacon (ca. 1214–ca. 1292), Bernard Cesi (Caesius) (ca. 1581–1630), Robert Boyle (1626–1691), Antoine Laurent Lavoisier (1743–1794). An attempt to honor one early investigator came with the introduction of the name avicennite in 1958, for the Persian physician Avicenna (Ibn Sina) (980–1037), who contributed much to medicine, physics (musical theory), chemistry, and mineralogy. At about the same time birunite, a mineral whose species status is still uncertain, was proposed to honor Abu al-Biruni (973–1048), an Arabic scientist of Uzbekistan, who determined specific gravities of minerals and metals.

MINERAL COLLECTORS

Mineral collectors and amateur mineralogists have been responsible for the discovery of many new species, especially in recent years. Because of their contributions to the furtherance of the science, it is fitting that they should be honored by mineral names, usually proposed by the professional who conducts the research on the new mineral.

Heulandite was named for John Henry Heuland (1778–1856) of England. A description of his collection was published in three volumes by Armand Lévy in 1837. Colonel Washington Augustus Roebling (1837–1926), the celebrated civil engineer and discerning mineral collector of Trenton, New Jersey, was honored by the mineral roeblingite. Morganite, a pink variety of beryl, was named for John Pierpont Morgan (1837–1913), the American banker and philanthropist, who also collected gems. Other established mineral species recently discovered by and named after amateurs and collectors include bjarebyite (Gunnar Bjareby, Boston, Massachusetts), brannockite (Kent C. Brannock, Kingsport, Tennessee), downeyite (Wayne F. Downey, Jr., Harrisburg, Pennsylvania), gerstmannite (Ewald Gerstmann, Franklin, New Jersey), jonesite (Francis T. Jones, Berkeley, California), perloffite (Louis Perloff, Tryon, North Carolina), yedlinite (Neal Yedlin, New Haven, Connecticut). A considerable number of the new minerals described in the *Mineralogical Record* honor collectors.

EXPLORERS

Minerals have been named in honor of explorers, not only for early adventurers but also for those more recent explorers of space and the Moon. Some of the more noteworthy examples of these include the following.

armalcolite, after the three American astronauts on the Apollo XI lunar mission in 1969: Neil Alden *Arm*strong (first man on the Moon), Edwin Eugene *Al*drin, Jr. (second man on the Moon), and Michael *Col*lins (who did not actually land).

armstrongite, Neil Alden Armstrong, American astronaut, first man on the Moon's surface, 1969.

chkalovite, Valery Pavlovich Chkalov, Russian aviator who made the first nonstop flight from Moscow via the North Pole to the United States, 1937.

coronadite, Francisco Vásquez de Coronado, the sixteenth-century Spanish explorer of the American Southwest.

gagarinite, Yuri Alekseevich Gagarin, Russian cosmonaut, and first man to travel in space, 1961.

humboldtine, Baron Alexander von Humboldt, early nineteenth-century German scientist and explorer, especially noted for his exploration of Latin America and Russia.

komarovite, Vladimir M. Komarov, Russian cosmonaut who was killed during his return from a space flight in 1967.

leifite, Leif Ericson, the tenth-century Norse mariner and explorer of Greenland and perhaps North America.

livingstonite, David Livingstone, the nineteenth-century Scottish missionary-doctor and African explorer.

OTHER PROFESSIONS

Although most mineral names are derived from persons usually identified as scientists, or related to mineral mining, collecting, or exploration, there are some cases where the person honored is well-known for contributions in other fields. For example, goethite is named after Johann Wolfgang von Goethe (1749–1832) the well-known German writer, poet, and philosopher. Goethe was also a keen student of natural sciences and made some interesting contributions to our knowledge of nature, including geology. Although the Swede Emanuel Swedenborg (1688–1772) is usually identified as a philosopher and mathematician, he made some important contributions to the theories of matter and crystal structure; swedenborgite was named for him. Clintonite was named in honor of the well-known American lawyer and statesman DeWitt Clinton (1769–1828), who was interested in geology and other sciences.

Two presidents of the United States have been honored by mineral names. In 1822 the name jeffersonite (Thomas Jefferson) was given to a pyroxene from Franklin, New Jersey. The mineral is no longer considered a separate species, but the name occasionally is used for varieties of augite and aegirine. Rooseveltite was named in honor of Franklin D. Roosevelt, the thirty-second president of the United States. Political leaders of other countries have likewise been honored by mineral names: rivadavite is for Bernardino Rivadavia, the first president of the Argentine Republic; holtite is for Harold Edward Holt, former Prime Minister of Australia; nyerereite is for Julius K. Nyerere, president of Tanzania.

Several names were derived from members of early European aristocratic families. Alexandrite (a variety of chrysoberyl) is named for Alexander II (1818–1881), czar of Russia from 1855 to 1881. Willemite is from William I (Willem Frederik) (1772–1843), king of the Netherlands from 1815 to 1840. Other minerals are named for barons, counts, dukes, and important military leaders. A few examples follow.

cancrinite, Count Georg Cancrin, former Russian Minister of Finance.

chevkinite, General K. V. Tschevkin, of Russia.

prehnite, Colonel Hendrik von Prehn, of Cape Town, South Africa.

samarskite, Colonel V. E. von Samarski-Bykhovets, of Russia.

stephanite, Archduke Victor Stephan, of Austria.

uvarovite, Count S. S. Uvarov, of Russia.

WOMEN'S NAMES IN MINERAL NAMES

Over the long history of mineral nomenclature, women's names have not been honored to the same degree as those of men. However, in recent years this trend has been changing and several well-known female mineralogists have been honored by mineral names.

One of the most interesting names is eveite, named for the mother of the human race, Eve. When the investigators of this mineral discovered its crystal-chemical relationship to the mineral adamite (named for Gilbert Joseph Adam), they thought the name eveite would be appropriate. The minerals form isostructural orthorhombic crystals. The formula for eveite is $Mn_2(AsO_4)(OH)$ while that for adamite is $Zn_2(AsO_4)(OH)$.

Laurite was named by Friedrich Wöhler, the German chemist, as a compliment to the wife (Laura) of his personal friend, Charles

Arad Joy, an American chemist. Eylettersite was named for the wife of the discoverer, L. Van Wambeke. Marialite is named after Maria Rosa, the wife of G. vom Rath, a German mineralogist.

Because the name curite had already been given to honor her husband Pierre Curie, Madame Marie Curie was honored with sklodowskite, derived from Sklodowska, her maiden name.

Other women recently honored for their contributions to the science of mineralogy are listed below.

bornemanite, Irina D. Borneman-Starynkevich, Russian mineralogist.

dellaite, Della Martin Roy, American mineralogist.

gaidonnayite, Gabrielle Donnay, American-Canadian mineralogist.

lonsdaleite, Kathleen Lonsdale, British crystallographer.

mroseite, Mary E. Mrose, American mineralogist.

schachnerite, Doris Schachner, German mineralogist.

shadlunite, Tatyana Shadlun, Russian mineralogist.

swinefordite, Ada Swineford, American clay mineralogist.

weeksite, Alice D. Weeks, American mineralogist.

PERSONAL NAMES HIDDEN IN MINERAL NAMES

In spite of the fact that most names derived from personal names are formed simply by adding an appropriate ending (usually -*ite* or -*lite*), there are some cases where the original name is considerably changed, especially in one of the following ways. First, a name may be altered by prefixing initials, first names, or parts of names to the family name. Second, the person honored may not be identified easily because only his first or middle name is used in the derivation. Third, there are hybrid words derived from bits and pieces of various names. Fourth, some personal names are translated into words of other languages from which the mineral name is then formed. A directory of various persons whose names are hidden in mineral names is given in Appendix II, but some examples of names modified by prefixing initials or parts of given names are given below.

afwillite, Alpheus F. Williams.

carlfriesite, Carl Fries.

francoanellite, Franco Anelli.

frankdicksonite, Frank W. Dickson.

gaidonnayite, Gabrielle Donnay.

haxonite, H. J. Axon.

jimthompsonite, James B. Thompson, Jr.

junitoite, Jun Ito.

tombarthite, Thomas F. W. Barth.

yofortierite, Y. O. Fortier.

The following names were derived from either the first or middle name of the person honored.

andorite, Andor von Semsey.

bellidoite, Eliodoro Bellido Bravo.

cliffordite, Clifford Frondel.

dellaite, Della M. Roy.

hulsite, Alfred Hulse Brooks.

leifite, Leif Ericson.

rustumite, Rustum Roy.

sterryite, Thomas Sterry Hunt.

A few mineral names have been formed by combining parts of the names of two or more people to form a new word (usually called a *portmanteau* word). Similar hybrid words have also been developed to form mineral names from earlier known minerals and from chemical components. Two examples of portmanteau names in which people are hidden are armalcolite (N. A. *Arm*strong, E. E.

*Al*drin, Jr., and M. *Col*lins) and ludlockite (F. *Lud*low Smith, III, and Charles *Lock*e Key).

A very oblique method of naming minerals after persons is applied in those few cases where, instead of using the name directly, the equivalent of the name in a different language is used. The name pavonite, for example, is derived from Latin *pavo, peacock,* and was used to honor Martin Alfred Peacock, the Canadian mineralogist. Swedish *stenhuggar,* which means *stonemason,* was used to form stenhuggarite in honor of Brian H. Mason, who contributed much to the study of Swedish minerals. The direct name masonite would not have been appropriate because the term is used as a trademark. Robert Mitchell Thompson, the Canadian mineralogist, was honored with the name twinnite. The name Thompson literally means "son of Thomas," and Thomas is derived from an Aramaic word meaning *a twin.* A name derived more directly from Thompson might have led to further confusion with names already in the nomenclature, thomsenolite and thomsonite. However, in another case, potential confusion with these names was avoided with the introduction of the name jimthompsonite.

MINERALS HONORING A FAMILY NAME
OR A PERSON MORE THAN ONCE

Although Smith is a very common name and one might predict only one Smith could be honored by having a mineral named for him, this is not the case. In fact several Smiths have been honored in this way. The most predictable and direct name is smithite (G. F. Herbert Smith). However, in addition to this there are joesmithite (Joseph V. Smith) and wardsmithite (Ward C. Smith). Although Smiths are honored by two additional minerals the word does not appear in the mineral names: lawrencite (John Lawrence Smith) and ludlockite (Frederick *Lud*low Smith, III, and Charles *Lock*e Key). Other names related to Smith are smithsonite (James Smithson) and smythite (Charles Henry Smyth, Jr.). Other names related to Smith are smithsonite (James Smithson) and smythite (Charles Henry Smyth, Jr.).

A similar situation is the one involving danalite and danaite. The first name is for James Dwight Dana while the second one, which is an obsolete variety term, was named for James Freeman Dana.

In some cases personal names have been used twice to honor the same individual. Examples of this follow.

berzeliite, berzelianite, Jöns Jacob Berzelius.

berthierite, berthierine, Pierre Berthier.

dufrenite, dufrenoysite, Pierre A. Dufrenoy.

fersmanite, fersmite, Alexander E. Fersman.

foshagite, foshallasite, William Frederick Foshag.

sudoite, tosudite, Toshio Sudo.

wolfeite, wroewolfeite, Caleb Wroe Wolfe.

In a few cases one person is honored once using his first and again using his last name.

andorite, semseyite, Andor von Semsey.

brianite, stenhuggarite, Brian H. Mason.

cliffordite, frondelite, Clifford Frondel.

esperite, larsenite, Esper S. Larsen, Jr.

MINERAL NAMES HONORING MORE THAN ONE PERSON

Several minerals have been given names which honor more than one person. These terms are either derived from family names, given names, or parts of names (portmanteau words).

Those derived from family names include the following.

ameghinite, Florentino Ameghino and Carlos Ameghino, brothers.

braggite, William H. Bragg and his son William Lawrence Bragg.

dahllite, Tellef Dahll and Johan M. Dahll, brothers.

franckeite, Carl Francke and Ernest Francke.

hauerite, Joseph von Hauer and Franz von Hauer.

iimoriite, Satoyasu Iimori and Takeo Iimori.

kupletskite, Boris M. Kupletsky and Elsa M. Bohnshtedt-Kupletskaya.

labuntsovite, Aleksander N. Labuntsov and Ekaterina E. Labuntsov-Kostyleva.

Some names have been given to honor entire families. Usually no individual member of the family is specified.

arnimite, von Arnim family of Zwickau, Germany.

kimzeyite, Kimzey family of Magnet Cove, Arkansas.

lemoynite, Charles Lemoyne and his four sons, of Canada.

rynersonite, Rynerson family of San Diego County, California.

sicklerite, Sickler family of Pala, California.

Arthurite was named to honor two men, but in this case the given names were used. The mineral is named for Arthur Russell and Arthur W. G. Kingsbury.

In a few cases parts of different family names have been joined to form mineral names honoring the various people involved. In addition to armalcolite and ludlockite, discussed above, the name *reedmergnerite* falls into this category (Frank S. *Reed* and John L. *Mergner*).

REFERENCES

1. Dana, Edward S. *System of Mineralogy of . . . Dana,* Sixth Ed. New York: John Wiley & Sons, 1892, p. xli.
2. Sage, B. G. "Observations sur la prehnite de M. Werner." *Journal de Physique, de Chimie, d'Histoire naturelle et des Arts* **34:** 446–449 (1789).
3. Estner, F. J. A. *Freymüthige Gedanken über Herrn Inspector Werners Verbesserungen in der Mineralogie, nebst einigen Bemerkungen über Herrn Assessor Karstens Beschreibung des vom sel. Leske hinterlassen Mineralien-Cabinetts.* Wien: I. Alberti, 1790.
4. Povarennykh, A. S. *Crystal Chemical Classification of Minerals,* Volumes I–II. New York: Plenum Press, 1972.
5. Fairbanks, E. E. "Joesmithite, etc." *Mineralogical Record* **7:** 146–147 (1976).

3
Names from Places

Names derived from localities are only second in importance to those named for persons. Nearly 500 such names apply to valid mineral species, some names having been used since ancient times, as magnetite for Magnesia, a land bordering ancient Macedonia. A large number of these are now obsolete, having been used as synonyms for other minerals. In some instances synonyms persisted in the literature as varietal names—for example, the varieties of diopside: alalite (Ala valley, Piemonte), traversellite (Traversella), canaanite (Canaan, Connecticut), mussite (Mussa Alp, Piemonte)—but with the passage of time they tend to fall into disuse.

Locality names can be placed in at least three major categories. The first, and by far the most common, consists of names after physiographic features of the Earth's surface: volcanos, mountains, hills, ridges, valleys, ravines, basins, plateaus, deserts, continents, peninsulas, capes, islands, fjords, seas, lakes, rivers, creeks, and springs. The second category consists of names derived from political or administrative divisions and tracts of land: countries, states, provinces, territories, counties, townships, quadrangles, towns, districts, and even farms. The third, and smallest, category includes names connected with the actual deposit in which the mineral was found; such terms may be from names of mines, shafts, quarries, pegmatites, and intrusions. A few illustrative examples follow.

NAMES AFTER PHYSIOGRAPHIC FEATURES

andesite, Andes Mountains.

antarctite, Antarctic Continent.

athabascaite, Lake Athabasca, Canada.

bityite, Mt. Bity, Madagascar.

cebollite, Cebolla Creek, Gunnison County, Colorado.

doloresite, Dolores River in Colorado.

elbaite, Island of Elba near Italy.

fayalite, Fayal Island in the Azores.

gamagarite, Gamagara ridge, Postmasburg district, South Africa.

garronite, Garron plateau area of Antrim, Ireland.

huanghoite, Huang Ho River in China.

ikaite, Ika fjord, near Ivigtut, Greenland.

ilmenite, Ilmen Mountains, Soviet Union.

inderite, Inder Lake, western Kazakhstan, Soviet Union.

jarosite, Jaroso ravine in Sierra Almagrera, Spain.

joaquinite, Joaquin ridge of the Diablo Mountain range, California.

kitkaite, Kitka River, Kuusamo, northeast Finland.

monetite, Island of Moneta, in the Caribbean Sea.

scawtite, Scawt Hill, County Antrim, Ireland.

vesuvianite, Mount Vesuvius, Italy.

wairauite, Wairau Valley, South Island, New Zealand.

yugawaralite, Yugawara hot spring, Kanagawa Prefecture, Japan.

NAMES AFTER POLITICAL DIVISIONS

Many names are after the countries where the mineral was first discovered. Columbite comes from an old name of the United States of America (Columbia). Other self explanatory terms include afghanite, brazilianite, cubanite, guyanaite, indialite, iranite, iraqite, malayaite, and surinamite. Recently a number of African nations have had minerals named after them: congolite, kenyaite, nigerite, tanzanite, tunisite, and zairite. American state names include californite, coloradoite, illite, iowaite, minnesotaite, montanite, oregonite, and texasite.

Miscellaneous names after political divisions are labradorite (Labrador Territory) and andalusite (Andalusia, a province of Spain). Several American county names are alleghanyite (Alleghany County, North Carolina), inyoite (Inyo County, California), sheridanite (Sheridan County, Wyoming), and sussexite (Sussex County, New Jersey). The following town and village names are among the many which have been incorporated in mineral names.

ajoite, Ajo, Arizona.

alamosite, Alamos, Sonora, Mexico.

aragonite, Aragon, Spain.

bararite, Barari, Bengal, India.

bytownite, Bytown, now Ottawa, Canada

chesterite, Chester, Vermont.

ettringite, Ettringen, Germany.

franklinite, Franklin, New Jersey.

grantsite, Grants, New Mexico.

nantokite, Nantoko, Chile.

terlinguaite, Terlingua, Texas.

tobermorite, Tobermory, Island of Mull, Scotland.

villamaninite, Villamanín, León Province, Spain.

NAMES AFTER MINES AND MINERAL DEPOSITS

A common practice in naming new minerals is to incorporate the name of the mine or deposit in which the mineral was first found. Some examples follow.

antlerite, Antler mine, Mohave County, Arizona.

bultfonteinite, Bultfontein diamond mine, Kimberley, South Africa.

carlinite, Carlin gold deposit, Eureka County, Nevada.

hummerite, Hummer mine, Montrose County, Colorado.

junoite, Juno ore body, Tennant Creek, Australia.

redledgeite, Red Ledge mine, Nevada County, California.

rosasite, Rosas mine, Sulcis, Sardinia.

starkeyite, Starkey mine, Madison County, Missouri.

susannite, Susanna mine, Leadhills, Scotland.

tintinaite, Tintina silver mines, Yukon Territory, Canada.

zunyite, Zuni mine, San Juan County, Colorado.

These are only a few of a relatively large number of names.

As a variation of the above, some terms are derived from the geological environment in which the mineral was discovered: muskoxite (Muskox intrusion, Northwest Territories, Canada), palermoite (Palermo pegmatite, North Groton, New Hampshire), tazheranite (Tazheran alkalic massif, west of Lake Baikal, Soviet Union), and walpurgite (Walpurgis vein, Weisser Hirsch mine, Schneeberg, Saxony).

SOME UNUSUAL EXAMPLES OF NAMES

An unusual nomenclatural pair are the minerals gladite and hammarite, both named for the same locality, at Gladhammar, Kalmar, Sweden.

Surely one of the most remarkable of all names is tranquillityite, a mineral discovered in rocks collected from the Sea of Tranquillity on the Moon during the Apollo XI mission.

Strange as it may seem, several minerals were named after incorrect localities. Milarite, for example, was first announced as having been found in Val Milar instead of the correct locality in Val Giuf, Switzerland. Some believe that Val Milar was deliberately given as the locality to deceive other mineral collectors. Doubt has also been expressed about the actual occurrence of linarite at Linares, Spain, from which place the mineral received its name. In still another example, adularia was named for the Adular Mountains in the Alps, which incorrectly were thought to be the range containing Mount St. Gotthard, the mineral's locality.

A few names combine the locality with other connotations. The term inderborite, for example, is after the locality at Inder Lake, Western Kazakhstan, Soviet Union and at the same time after the fact that the mineral is a borate. Uralborite was named in similar fashion from its occurrence in the Urals and its chemical composition. Ilmenorutile is named after the Ilmen Mountains, where it was first found, and its chemical relationship to rutile. In like manner, ulvöspinel is named for the locality at Södra Ulvön, in northern Sweden, and its relationship to spinel. Another term which combines the locality with an additional feature is microsommite. This name combines the Greek prefix for *small,* in allusion to its minute prismatic crystals, and the locality at Monte Somma, Vesuvius, Italy, where it was first discovered.

At first glance one might believe that strontianite was named for the chemical element strontium. However, the mineral was first named after Strontian in Argyllshire, Scotland, and it was only later that it was discovered that a new element was present in this mineral; the element, when isolated, was named strontium.

USE OF OBSOLETE GEOGRAPHICAL NAMES

Several mineral names incorporate obsolete geographical names. Many of these are in Latin or other ancient languages.

ardealite, from Ardeal, the old Rumanian name for Transylvania where the mineral was initially found.

caledonite, from Caledonia, an ancient name for Scotland.

cornubite, from Cornubia, the Latin name for Cornwall, England.

cymrite, from Cymru, the Welsh name for Wales.

hemusite, from Hémus, an ancient name of the Balkan Mountains where the mineral occurs.

ilvaite, from the Latin name for Elba, the island on which the mineral was found.

ruthenium, from *Ruthenia,* the Latin name for Russia, the locality of the first specimens.

sogdianite, from Sogdiana, the name of an ancient country of central Asia comprising part of modern Turkestan.

sursassite, derived from the Roman name Sursass for Oberhalbstein, Switzerland.

variscite, from Variscia, an ancient name of the Voigtland district of Germany, where the mineral was first found.

4

Impersonal Names

Although the number of impersonal mineral names is relatively small, their origins are interesting, as may be readily appreciated from the several categories given below.

NAMES DERIVED FROM OTHER MINERAL NAMES

Some minerals have been named for earlier known minerals or chemical substances because they resemble the substances in physical properties or bear chemical or crystallographic similarities.

Some descriptive terms were formed using the common suffix *-oid,* which connotes *like:* for example, the mineral chloritoid (*chlorite-like*) and the groups feldspathoid (*feldspar-like*) and pyroxenoid (*pyroxene-like*). The mineral names stannoidite and triploidite are similarly derived from the resemblances of these minerals to stannite and triplite, respectively. Demantoid was so named because its high luster resembles that of the diamond (German, *Demant*).

In another class of names, the suffix *-ite,* meaning *of the nature of* or *similar to,* has been added to well-established mineral names to form new names for minerals, or mineral varieties, which possess similar characteristics. Some examples, are jadeite (*similar to jade*) and schorlomite (*similar to* or *the same as schorl*). The suffix *-ine,* meaning *like,* also is used to form such names as sapphirine (*like sapphire*). Occasionally a name was derived from a well-known chemical substance, rather than an earlier mineral, as in phosgenite (*similar to phosgene*) and portlandite (from Portland cement, in which the mineral was first noted).

Nekoite is a unique mineral name in that it was derived from okenite, a species named for Lorenz Oken, by reversing the order of the letters of Oken's name; hence *oken* becomes *neko*. Nekoite was originally mistaken for okenite because of its similar composition.

Some minerals and mineral group names have been formed by combining parts of existing words to form portmanteau words. The name maghemite was formed by combining parts of the words *mag*netite and *hem*atite. Several mineral group or subgroup names originated in the same way: kandite (*ka*olinite, *n*acrite, *d*ickite), biopyribole (*bio*tite, *pyr*oxene, amph*ibole*), pyralspite (*pyr*ope, *al*mandine, *sp*essartine), ugrandite (*u*varovite, *gr*ossular, *and*radite). Such words serve as mnemonic aids for remembering the members of mineral groups, but so far they have not received total acceptance.

The largest number of names derived from earlier mineral names consists of those formed by adding appropriate prefixes, such as those listed in Chapter 9.

The Greek prefix *para-* is found attached to many preexisting mineral names. Generally this prefix means *beside* or *near* and the new mineral so designated has a close relationship to or can be compared with the earlier mineral. Often it denotes a dimorphous (two crystal structures) relationship between two compounds, as in parabutlerite (orthorhombic)/butlerite (monoclinic), parahilgardite (triclinic)/hilgardite (monoclinic), and parajamesonite (orthorhombic)/jamesonite (monoclinic). It may also indicate an association of the two minerals in occurrence, as with parakeldyshite and keldyshite, and paramontroseite and montroseite.

Another common prefix, *meta-*, which literally means *along with*, is frequently used to name a mineral which has a composition nearly identical to the one which is modified but contains less water. This is illustrated by the pairs rossite ($CaV_2O_6 \cdot 4H_2O$) and metarossite ($CaV_2O_6 \cdot 2H_2O$), and autunite ($Ca(UO_2)_2(PO_4)_2 \cdot 10$–$12H_2O$) and meta-autunite ($Ca(UO_2)_2(PO_4)_2 \cdot 2$–$6H_2O$). In metacinnabar the prefix *meta* means *with*, because the original specimens were associated with cinnabar.

Chemical prefixes are commonly used in mineral names to denote structural relationships but some chemical differences between the minerals. Mineral isotypes, which are similar or identical in crystal structure but differ slightly in chemical composition, are often named in this way. The jarosite group is a good example: normal jarosite is $KFe_3(SO_4)_2(OH)_6$, but silver-bearing argentojarosite is $AgFe_3(SO_4)_2(OH)_6$, while sodium-bearing natrojarosite is $NaFe_3(SO_4)_2(OH)_6$.

Where substances with the same chemical composition crystallize in more than one structure (polymorphs, or if two, dimorphs), prefixes denoting crystal symmetry have been used. Clinoenstatite is identical to enstatite in composition, but enstatite is orthorhombic whereas its dimorph clinoenstatite is monoclinic. Similarly, orthochamosite is the orthorhombic dimorph of chamosite which is monoclinic; tetrawickmanite is a tetragonal dimorph of wickmanite which is isometric.

NAMES DERIVED FROM BOTANICAL TERMS

Some of the most interesting mineral names are derived from real or imagined resemblances to plants, perhaps mode of growth, color, a seed, a fruit, or other attributes. Important examples include the following:

anthophyllite, from Neo-Latin *anthophyllum* (= *clove*), in allusion to its clove-brown color.

ericaite, named for its violet color, which resembles that of the heather blossom (genus *Erica*).

galaxite, from the herb galax, which is abundant in the district of North Carolina where the mineral was discovered. The nearby town of Galax, Virginia, was also named after the plant.

garnet, named from Latin *granatum* (= *pomegranate*), the seeds of which it was thought to resemble.

grossular, from Latin *grossularium* (= *gooseberry*), alluding to the pale green color of some varieties.

hyacinth, from Greek for the member of the lily family of the same name, because of its hyacinth-red color.

malachite, from Greek for *mallow,* in allusion to the green color of the mineral, which resembles that of the mallow leaf.

NAMES FROM INSTITUTIONS AND EXPEDITIONS

Because of their affiliation with various institutions or scientific expeditions, some mineralogists chose to name newly discovered minerals in their honor.

azoproite, from the Russian title for the International Geological Association for the Study of Deep Zones of the Earth's Crust (AZOPRO), because it was found during the preparation of a guidebook for the Association's meeting at Lake Baikal in 1969.

carletonite, for the institution where it was first identified, Carleton University, Ottawa, Canada.

carlsbergite, from the Carlsberg Foundation, Copenhagen, which supported research on the mineral.

halurgite, from the Institute of Halurgy of the Soviet Union, which has studied saline deposits for many years.

imgreite, from IMGRE, initials for the Russian name of the Institute of the Mineralogy, Geochemistry, and Crystal Chemistry of Rare Elements, in the Soviet Union.

nimite, from the initials NIM for the National Institute of Metallurgy, South Africa.

romarchite, and the related **hydroromarchite,** from the initials of Royal Ontario Museum (ROM), Toronto, Canada, and *arch*aeology, a department within the museum; these minerals were discovered and described by scientists there.

stevensite, from the Stevens Institute of Technology, Hoboken, New Jersey.

vimsite, from the initials VIMS of the Russian name for the All-Union Research Institute of Mineral Resources of the Soviet Union.

Mineral names which honor scientific expeditions include amakinite and raite. The Amakin Expedition prospected the diamond deposits in Yakutia, Soviet Union. Raite was named in honor of the international team of scientists who took a voyage (1969–1970) on the papyrus ship *Ra,* under the leadership of Thor Heyerdahl.

NAMES FROM COMPANIES

Several companies, especially those involved in mining, have been honored in mineral names.

amosite, from AMOS, the initial letters of Asbestos Mines of South Africa, where the mineral was discovered.

falcondoite, for the company which mines laterite at the type locality (near Bonao, Dominican Republic), Falconbridge Dominicana C. por A. (Falcondo).

lillianite, for the Lillian Mining Company, which operated mines near Leadville, Colorado.

santafeite, from the Atchison, Topeka, and Santa Fe Railroad Company, in recognition of its pioneer exploration and development of the uranium deposits of New Mexico.

tsumcorite, from the Tsumeb Corporation; the mineral was found in the Tsumeb mine, South West Africa.

NAMES FROM AMERICAN INDIAN TRIBES

Although they are not numerous, some mineral names of recent origin have been derived from Indian tribes of the Americas. Incaite was named for the Incas, who were the first recorded miners at the type locality of the mineral in Bolivia. Navajoite was named for its occurrence on the Navajo Indian Reservation in Apache County, Arizona. The Papago Indian tribe, which once inhabited the region around the mining center of Ajo, Arizona, has its name preserved in papagoite. Yavapaiite is named for the Yavapai tribe, which inhabited the area of Arizona around Jerome, the type locality. Although they are not tribal names, Mexican Indian terms are found in the names quetzalcoatlite (god of sea, agriculture, etc.), tlalocite (god of rain), and xocomecatlite (from the Nahua word for grapes).

NAMES FROM MYTHOLOGICAL CHARACTERS

Several mineral names are derived from the names of mythological characters. Generally these terms allude to some property of the mineral, or in some cases to relationships between pairs of minerals. Because these names are rather unusual some of them are discussed here.

aegirine, for Ægir, the Scandinavian god of the sea, the mineral being reported first from Norway.

atheneite, for the Greek goddess Pallas Athena, the virgin goddess of wisdom, war, arts, and crafts, in reference to the mineral's palladium content (see below).

ixiolite, from Ixion, a Thessalian king related to Tantalus, because of its relation to tantalite.

mercury, from the Roman god Mercury, son of Jupiter and Maia, a messenger and interpreter of the gods; named in allusion to its mobility.

neptunite, for the Roman god of the sea, Neptune; when discovered, this mineral was associated with aegirine, which had been named for Ægir, the Scandinavian god of the sea.

orpheite, for Orpheus, legendary Greek poet and musician, because of the association of Orpheus with the type locality of the mineral in the Rhodope Mountains, Bulgaria.

palladium, after the asteroid Pallas, which was discovered at about the same time; the name comes from the Greek goddess Pallas Athena.

pollucite, from Pollux, in Classical mythology the twin brother of Castor, for its association on the island of Elba with the mineral castor (obsolete name for petalite).

quetzalcoatlite, from Quetzalcoatl, a god of the Toltec and Aztec Indians of Mexico, who is identified with the sea; named for the relationship between its blue color and the sea.

tantalite, from Tantalus, a king of Greek mythology who was condemned to stand up to his chin in water under branches laden with fruit, the water or fruit receding whenever he tried to eat or drink; named in allusion to the tantalizing difficulties encountered in dissolving this mineral in acids preparatory to analysis.

tapiolite, for the Finnish god of the forest, Tapio; the mineral was discovered in Finland.

tlalocite, from the ancient Mexican god of rain, Tlaloc, in allusion to its high water content.

vertumnite, for Vertumnus, ancient Etruscan god, worshiped as the divinity of the changing seasons and growing plants. The mineral was found in the region of Italy where the god was worshiped.

5
Names from Chemical Composition

In contrast to most mineral names discussed so far, which say very little or nothing about the nature of the mineral itself, some names have been derived from the chemical components of the minerals. Although practically none of these reveal the complete chemical composition they do indicate the more important elements and occasionally give additional information.

NAMES FROM CHEMICAL ELEMENT NAMES

Numerous names are derived from, or are closely related to, the standard designations of the chemical elements; some of these are: aluminite, bismite, boracite, calcite, cerianite, chromite, cobaltite, gallite, germanite, hafnon, indite, lanthanite, magnesite, manganite, molybdenite, sodalite, titanite, tungstite, uraninite, vanadinite, yttrialite, and zincite. However, one cannot always be certain that the name was derived from the element, because there are cases where the element was named after the mineral: beryllium (from beryl), fluorine (from fluorspar or fluorite), nickel (from kupfernickel or nickeline), samarium (from samarskite), and zirconium (from zircon).

In addition to mineral names which correspond rather closely with standard modern names for chemical elements, other minerals are derived from old Latin or Greek terms for elements. Some examples from Latin include argentite (*argentum* = silver) and stannite (*stannum* = tin); some are used in combinations like plumboferrite (*plumbum* = lead; *ferrum* = iron) and auricupride (*aurum* = gold; *cuprum* = copper). Examples from Greek include chalcocite (*chalkos* = copper) and siderite (*sideros* = iron). Although they do not reveal the chemical components of the minerals, there are some names derived from Greek which refer to various chemical properties. These are considered in Chapter 6.

CHEMICAL PREFIXES IN MINERAL NAMES

Numerous chemical prefixes are used in mineral species names. Prefixes such as *arseno-, bismuto-, calcio-, chalco-, cupro-, ferri-, ferro-, kali-, lithio-, magnesio-, mangano-, natro-, phospho-, plumbo-, sidero-, thoro-, urano-,* and others, are very common. Frequently these are affixed to earlier mineral names to show structural and chemical relationships between the minerals. Sometimes the prefixes are combined with other Latin or Greek terms to form new names. Names like kaliophilite and natrophilite are formed from a prefix plus *phil,* which means *friend of,* hence: *friend of* potassium (*kalio-*) and *friend of* sodium (*natro-*) respectively. Chalcophyllite, from *phyllon* meaning *leaf* plus *chalco* for copper, giving *copper leaf,* alludes to the mineral's composition and micaceous structure; chalcotrichite in similar fashion means, literally, *copper hair,* alluding to the mineral's composition and exceedingly long, hairlike crystals. Siderophyllite, *iron leaf,* and siderotil, *iron fiber,* have similar meanings.

A complete list of chemical prefixes is given in Chapter 9.

Most varieties of mineral species are a result of relatively minor chemical deviations from the ideal composition of the mineral. To indicate such varieties a chemical adjective may be used to modify the mineral species name. Using this scheme such variety names as the following have been used: ferroan dolomite, mercurian tetrahedrite, zincian spinel, cobaltoan siderite, potassian natroalunite, aluminian chalcosiderite. In the last three examples the chemical adjectives modify mineral names which are themselves derived from common chemical prefixes. Palache, Berman and Frondel [1] discuss the use of chemical adjectives, originally proposed by Schaller,[2] in the naming of mineral varieties. They also present a complete list of the chemical adjectives which can be used in this manner, and discuss how different valencies can also be indicated.

CHEMICAL PORTMANTEAU WORDS

To include more information about the chemical compositions of minerals within their names, some mineralogists formed words by combining parts of the names of chemical elements, that is, portmanteau words. In some cases these are combinations of chemical symbols and parts of either the international or Greek or Latin element names. Some interesting examples follow.

armangite, *ar*senic and *mang*anese.

alvanite, *al*uminum and *van*adium.

aurostibite, gold (*aur*um) and antimony (*stib*ium).

calcurmolite, *calc*ium, *ur*anium, and *mol*ybdenum.

ferdisilicite, iron (*fer*rum), two (*di*), and *silic*on.

mourite, *mo*lybdenum and *ur*anium.

phosinaite, *pho*sphate, *s*ilicon, and sodium (*na*trium).

plumalsite, lead (*plum*bum), *al*uminum, and *s*ilicon.

ruthenarsenite, *ruthen*ium *arsen*ide.

silhydrite, a *sil*ica *hydr*ate.

sulvanite, *sul*fur and *van*adium.

telargpalite, *tel*lurium, silver (*arg*entum), and *pal*ladium.

uvanite, *u*ranium and *van*adium.

vanoxite, *van*adium *ox*ide.

zirsinalite, *zir*conium, *si*licon, and sodium (*na*trium).

In a few cases the crystal system has also been incorporated within the portmanteau word, as in the following.

hexatestibiopanickelite, *hexa*gonal, *te*llurium, antimony (*stib*ium), *pal-*ladium, *nickel.*

isoferroplatinum, *iso*metric, iron (*ferr*um), *platinum.*

tetraferroplatinum, *tetra*gonal, iron (*ferr*um), *platinum.*

NAMES DERIVED FROM CHEMICAL SYMBOLS

Other mineral names have been formed by using a combination of the standard chemical symbols for some or for all the elements in the mineral.

asbecasite, As, Be, Ca, Si.

babefphite, Ba, Be, F, P, H.

behoite, Be, H, O.

cafetite, Ca, Fe, Ti.

fenaksite, Fe, Na, K, Si.

kalsilite, K, Al, Si.

natisite, Na, Ti, Si.

rauvite, Ra, U, V.

sinoite, Si, N, O.

tinaksite, Ti, Na, K, Si.

umohoite, U, Mo, H, O.

In some rare cases these acronym names are modified to include additional data. Three polymorphs of $KAlSiO_4$ have been named kalsilite, trikalsilite, and tetrakalsilite. In trikalsilite the crystallographic *a* axis is three (*tri-*) times longer than that for kalsilite; in tetrakalsilite the same axis is four (*tetra-*) times longer.

PROPOSED USE OF CHEMICAL NAMES

Povarennykh,[3] a modern critic of mineralogical nomenclature, supports the trend in mineralogy which uses names expressing chemical compositions and other properties. However, he feels the methods so far employed have been insufficient. He rightly states that in "a sound scientific nomenclature, the close relation of name to nature enables us to deduce from the name the place of the object in the system." In his discussion of a new rational nomenclature for minerals he suggests, first, that each name should express all the elements contained in the mineral in the same sequence as in the chemical formula. Second, a suffix characterizing the electronegative element or radical, which determines the chemical class, should be used: for example, for sulfides, -*sulite;* for silicates, -*silite;* for oxides, -*oxite;* and so on. A third feature of the name might be an indication of the structure of the mineral, for example -*phyllite* (Greek *phyllon = leaf*), -*tilite* (Greek root *til-* = *fiber*), and -*tectite* (for *framework,* after Greek *tektōn = carpenter*). Using his scheme the mineral now known as greigite (named for J. W. Greig), $FeFe_2S_4$, would become ferdiferrisulite, while smythite (named for C. H. Smyth, Jr.), which is identical in composition, but has a layered or sheet structure, would be called ferdiferrisulphyllite. Other aspects, relating to water of hydration, polymorphism, polytypism, etc., could also be incorporated, and each of these is fully considered in his discussion of proposed mineral names. The use of Arabic numerals in mineral names has also been suggested in those cases where minerals have the same qualitative compositions, but differ quantitatively: for example, plum5stib8sulite for plagionite ($Pb_5Sb_8S_{17}$) and plum7stib8sulite for heteromorphite ($Pb_7Sb_8S_{19}$). Although these would be awkward names, one has to admit they convey much information about the chemistry of minerals. This concept is further discussed in Chapter

11. Povarennykh [3] has published his proposed mineral names, alongside the common names, in his two-volume work *Crystal Chemical Classification of Minerals*. These proposed names have not been accepted by the mineralogical community, which is strongly opposed to unilateral decisions regarding major revisions of nomenclature. The Commission on New Minerals and Mineral Names of the International Mineralogical Association privately circulates suggestions of this nature, and approves their publication only after there is agreement on major changes. In this way, introduction of numerous conflicting systems can be prevented.

REFERENCES

1. Palache, C., Berman, H., and Frondel, C. *System of Mineralogy of . . . Dana*, Seventh Ed., Volume I. New York: John Wiley & Sons, 1944.
2. Schaller, W. T. "Adjectival ending of chemical elements used as modifiers to mineral names." *American Mineralogist* **15:** 566–574 (1930).
3. Povarennykh, A. S. *Crystal Chemical Classification of Minerals*, Volumes I–II. New York: Plenum Press, 1972, p. 87.

6
Names from Greek Terms: Primarily Incorporating Physical and Chemical Properties

Traditionally names conveying information about crystallographic and other physical properties of minerals are derived from the Greek language. Features such as crystal systems, crystal habits, twinning, cleavage, colors, lusters, transparency, and many other properties, are incorporated in these Greek-derived names. In some cases these names also convey information about chemical properties. To a lesser extent the same was done by use of Latin terms.

CRYSTALLOGRAPHIC NAMES

Typically such terms are derived from Greek words denoting symmetry, crystal form or shape, unusual habits, or the crystal system.

Names derived from crystal forms include: tetrahedrite (tetrahedron), octahedrite (octahedron—obsolete name for anatase), rhomboclase (rhomboid crystals showing cleavage), and trigonite (trigonal shape).

Very numerous are names signifying crystal shapes or habits, although the crystal symmetry is not usually evident from the terms.

acanthite, *thorn.*

acmite, *point.*

anatase, *extension,* in allusion to the steeper bipyramidal form as compared to a true octahedron.

axinite, *ax,* wedgelike shape.

cylindrite, in allusion to the cylindrical form of the mineral.

diopside, *double appearance,* because the vertical prism zone can be oriented in two ways.

epidote, *increase,* since the base of the rhomboidal prism has one side longer than the other.

hemimorphite, from the hemimorphic symmetry of the crystals.

kentrolite, *spike,* in allusion to its prismatic habit.

microlite, *small,* because the type crystals were very tiny.

pinakiolite, *small tablet,* for its thin tabular habit.

plagionite, *oblique,* for monoclinic crystals.

platynite, *to broaden,* for platelike structure.

sanidine, *tablet,* in allusion to the tabular habit.

scapolite, *shaft,* in allusion to the prismatic habit.

The crystal system of a mineral may also be indicated by the use of Greek prefixes. From the isometric or cubic system, *iso-* may be used to form names like isoferroplatinum. The tetragonal system may be indicated by *tetra-,* as in tetraferroplatinum, tetranatrolite, and tetrawickmanite. In the hexagonal system (*hexa-*) there are hexatestibiopanickelite and hexagonite (originally supposed to be hexagonal, although actually monoclinic). The prefix *ortho-* has been used frequently to indicate orthorhombic symmetry, especially in the case of those substances which are polymorphous, that is, where both orthorhombic and monoclinic or other structures exist; for example, orthoantigorite, orthoferrosilite, and orthopinakiolite. In the same fashion, monoclinic minerals may be indicated by *clino-*, as with clinohumite, clinohypersthene, and clinozoisite. A simple prefix has not been used to indicate triclinic minerals, although terms implying this system may be seen in anorthite (meaning *not straight*) and plagioclase (meaning *oblique fracture*).

NAMES BASED ON CRYSTAL TWINNING

Numerous minerals are characterized by unique twinning. Twinning is where two or more crystals of the same species are grown together in a nonparallel, but a symmetrical and rational fashion. Several minerals have been named for their characteristic twinning: tridymite (*threefold twin*), tetradymite (*fourfold twin*), pentagonite (*five-angle* or fivelings). Staurolite is derived from a word meaning *cross,* in allusion to its cruciform twins. Polydymite refers to *many twins,* and was named because the mineral is often observed in twinned forms. Contrary to what one might at first expect, twinnite was not named for crystal twinning; it is named for Robert M. Thompson, as discussed in Chapter 2.

NAMES BASED ON MINERAL CLEAVAGE

Several minerals in the feldspar group owe their names to cleavage. An examination of these feldspar names will show the common recurrence of the suffix *-clase* (*fracture*). Orthoclase (*straight fracture*) has a right angle (90°) between its two best cleavage directions, while microcline (*little slant*) has a slight variation of the cleavage angle from 90°. When this cleavage angle departs even more from 90° we find plagioclase (*oblique fracture*) as the series name. Further, oligoclase (*little fracture*), a member of the plagioclase series, was so named since it was believed to have a less perfect cleavage than albite, another member of the series. Anorthoclase (*not-straight fracture*) also has a cleavage angle only slightly deviating from 90°.

Several other characteristically cleavable minerals also incorporate the root -*clase* in their names: alloclasite (*another fracture*), clinoclase (*inclined fracture*), isoclasite (*equal fracture*), periclase (*around fracture*, meaning cubical cleavage), rhomboclase (from rhomboid crystal habit and basal cleavage).

Still other mineral names based on cleavage include the following: amblygonite (*blunt angle* between cleavages), petalite (*leaf*, for leaflike or planar cleavage), triplite (*threefold*, apparently in reference to three cleavages). The prefix *lepido-*, meaning *scale* or *flake*, has also been used to form names for minerals which have good cleavage in one direction, usually a basal cleavage, for example, lepidolite (*scale*) and lepidomelane (*scale* and *black*).

NAMES DERIVED FROM MINERAL COLORS

The Greek terms used in the derivation of these names are either from color designations or from names of substances which are noted for their characteristic colors. Examples of this method of naming follow.

achroite, *no color* or *colorless.*

amarantite, *amaranth red.*

attakolite, *salmon red.*

carpholite, *straw yellow.*

chlorite, *green.*

crocoite, *saffron orange.*

erythrite, *red.*

glaucochroite, *blue color.*

ianthinite, *violet.*

kyanite, *blue.*

leucite, *white.*

malachite, *mallow green.*

melanite, *black.*

phlogopite, *firelike,* for reddish tinge.

phoenicochroite, *deep red color.*

pyrope, *firelike,* from color.

rhodonite, *rose.*

sarcolite, *flesh red.*

spodumene, *ash gray.*

xanthiosite, *yellow* and *sulfur.*

NAMES DERIVED FROM MINERAL LUSTERS

Because the luster of a mineral is a rather characteristic property, it is not surprising to find names indicating this property.

augelite, *luster,* from its pearly luster.

augite, *luster,* from its appearance in dark rocks.

chiolite, *snow,* from its appearance and similarity to cryolite.

cryolite, *ice,* from its appearance.

eleolite, *oil,* from its greasy luster.

ganomalite, *luster,* having a high splendent luster.

ganophyllite, *luster* and *leaf,* alluding to the high luster of its cleavage plates.

margarite, *a pearl,* alluding to its luster.

pachnolite, *frost,* alluding to its appearance.

pimelite, *fatness,* from its greasy appearance.

sericite, *silky,* in allusion to its appearance.

NAMES REFERRING TO OTHER PHYSICAL PROPERTIES AND ASSOCIATIONS

Mineral transparency also has been considered in naming some minerals. Only a few examples are given here.

adelite, *obscure,* in allusion to its lack of transparency.

dioptase, *through* and *to see,* for its transparency.

hyalophane, *glass* and *to appear,* in allusion to its transparency and appearance.

Additional physical properties, used in the formation of mineral names from Greek, are illustrated by the following.

allactite, *to change,* in allusion to its strong pleochroism.

analcime, *weak,* in allusion to its weak electrical properties when heated or rubbed.

barite, *heavy,* from its high specific gravity.

clinochlore, *to incline* and *green,* from the obliqueness between the optical axes and its green color.

diamond, a corruption of Greek *adamas,* meaning *invincible,* in allusion to its hardness and durability.

pyrite, *fire,* because it is a substance which strikes sparks with steel.

psilomelane, *smooth* and *black,* from its feel and color.

xanthoconite, *yellow* and *powder,* because of its yellow streak.

Still other names derived from Greek roots relate to similarities between two minerals, or to some special occurrence of the mineral in nature. Some of these names are included below.

corundophilite, from the mineral corundum and Greek for *friend,* because it is associated with corundum.

elpidite, *hope,* because there was a reasonable expectancy of finding other interesting minerals from the same locality.

emplectite, *entwined,* because it was associated with quartz.

neotocite, *newborn,* alluding to its recent origin; an alteration product of rhodonite.

phenakite, *a cheat,* because it had been mistaken for quartz.

pyroxene, *fire* and *a stranger,* originally erroneously thought to be out of place in igneous rocks.

tychite, *luck* or *chance,* in allusion to the circumstance that of two crystals discovered in a stock of about 5000 northupite crystals, the first and one of the last ten proved to be tychite.

NAMES RELATING TO CHEMICAL BEHAVIOR

There are numerous additional names, derived from Greek terms, which imply certain chemical behavior or chemical properties of minerals. Because the blowpipe was used so extensively in testing in the earlier days of mineralogy, many are related to such use.

aeschynite, *shame,* alluding to the fact that chemistry at that time had no means for separating titanic acid and zirconia.

allophane, *another* and *to appear,* because it changes before a blowpipe.

anhydrite, *without water,* in contrast to the similar hydrated mineral gypsum.

apophyllite, *away from* and *leaf,* from its tendency to exfoliate when heated.

cacoxenite, *a bad guest,* because the mineral injures the quality of iron made from ore in which it is found.

dixenite, *two* and *stranger,* from the association of silica and arsenious oxide in the same compound.

enstatite, *an opponent,* because it is so refractory before the blowpipe.

eucrasite, *well-mixed,* on account of its complex composition.

eudialyte, *easy to undo,* on account of its easy solubility in acids.

euxenite, *friendly to strangers,* because it contains so many rare ingredients.

griphite, *puzzle,* in allusion to its chemical composition.

mesolite, *middle,* chemically between natrolite and scolecite.

nepheline, *a cloud,* because in strong acids it becomes cloudy.

pharmacolite, *poison,* because of its arsenic content.

picromerite, *bitter* and *a part,* in allusion to its content of magnesia (salts of which are often bitter).

pyrolusite, *fire* and *to wash,* used to free glass of colors by its oxidizing effects on iron.

rhabdophane, *a rod* and *to appear,* in allusion to the characteristic bands shown in its chemical spectrum.

rhodizite, *to be rose-colored,* because it tinges the blowpipe flame red.

tachyhydrite, *quick* and *water,* because it deliquesces so easily.

Other prefixes derived from Greek and used to indicate the chemical elements are incorporated in mineral names. These are discussed elsewhere, especially in Chapter 5 and Chapter 9.

ANCIENT GREEK MINERAL NAMES

Many names derived from Greek words have already been presented in this chapter, but it should be noted that most of these were coined in recent centuries. However, some names still in common use were also used by the ancient Greeks. Spellings in some instances have been modified somewhat, but generally the terms go back to antiquity. The origins of some of these words are uncertain.

agate, from *achates,* occurred in the river Achates, perhaps the modern Carabi or the Cannitello, in southwestern Sicily; the ancient term included some materials now referred to by other names.

alabaster, from *alabastrites,* the stone out of which a vase called an *alabastron* was made; originally applied to onyx marble (calcite) as well as gypsum.

amethyst, Greek for *not drunk,* was believed to have power to counteract drunkenness.

asbestos, Greek for *inextinguishable,* from the misconception that if the material were ignited the fire could not be put out.

beryl, from *beryllos,* whose origin is not known, but was used by the Greeks for more than one kind of green gem; it has been suggested that the German word *Brille,* for eyeglasses, and *beryl* are related and derived from some now obscure ancient word.

chrysocolla, Greek for *gold* and *glue,* a name given by the ancients to a mineral or minerals used for soldering gold, but long applied to various green copper minerals.

emerald, from Greek *smaragdos;* was applied by the ancients to various stones of pronounced green color.

gypsum, from *gypsos* (= *plaster*), an ancient name which applied equally to gypsum, dehydrated gypsum, and lime.

hematite, from *haimatitis* (= *blood-red*); used for the same mineral today as in ancient times, though it may also have referred to jasper and other red materials.

7

Names from Latin Terms

Although they are not as numerous as Greek names, one will find mineral terms derived from Latin as well. Generally these terms are either related to the physical properties of the minerals or to their chemical compositions.

NAMES RELATED TO PHYSICAL PROPERTIES

Among names associated with physical properties color terms are used most frequently. The terms used are either color designations or are derived from things which have characteristic colors. Names based on colors include:

albite, white.

aquamarine, *seawater,* for its seagreen color.

carbuncle, from *carbunculus* (= *little coal*), for its firelike color.

carnelian, *fleshy,* fleshlike color.

celestite, *celestial,* light, heavenly blue.

corvusite, *raven,* blue black.

indicolite, indigo.

lavendulan, lavender.

purpurite, purple.

rubellite, reddish.

ruby, red.

rutile, red.

violarite, violet.

The lusters of a few minerals are expressed by names derived from Latin. For example, mica apparently comes from *micare,* meaning *to shine.* Plumbogummite, *lead* and *gum,* is named for its composition and gumlike luster. Other names allude to the general shape or form of the specimens. For example, cuspidine, *a spear;* garnet (*granatum = pomegranate*), from its resemblance to the seeds of a pomegranate; and vermiculite, *to breed worms,* because when heated it exfoliates into wormlike threads. Other names derived from various Latin words are: fluorite, *to flow,* because it melts easily; ruthenium, *Russia,* the country of its occurrence; and tellurium, *the Earth.*

CHEMICAL NAMES FROM LATIN

Probably the greatest influence Latin has had in forming mineral names is in chemical prefixes. Mineral names involving the following are at least in part of Latin origin: *alumino-, argento-, auri-, calcio-, carbo-, cupro-, ferri-, ferro-, fluor-, nitro-, plumbo-,* and *stanno-.* The use of these prefixes, as well as similar Greek prefixes, to modify older mineral names, or to form new mineral names is discussed in Chapter 5.

Some examples of mineral names derived directly from Latin designations for the common metallic elements are argentite (for silver), cuprite (for copper), and stannite (for tin).

NAMES FORMED FROM A COMBINATION
OF LATIN AND GREEK TERMS

Mineral names made half of Latin and half of Greek terms are usually considered objectionable. Good examples of mineral names formed this way are few. These hybrid names can be illustrated by allargentum, from Greek for *another* and Latin for *silver,* and pyroaurite, from Greek for *fire* and Latin for *gold.* A less objectionable situation involves the many cases where names are formed by combining chemical element terms and prefixes with older mineral names of various origins. The following names, for example, illustrate the use of Latin prefixes with terms derived from Greek: ferrimolybdite, ferroaxinite, ferrocarpholite, plumbopyrochlore, and stannopalladinite. Some cases involve the combination of words derived from other languages with Latin or Greek terms: tetraferroplatinum (Greek, Latin, and Spanish), zincobotryogen (German and Greek), and hexatestibiopanickelite (Greek, Latin, and German).

8
Names from Various Languages

Although mineral names based on terms from the Greek and Latin languages far outnumber names derived from other tongues, one should not minimize the importance of the contributions of these other languages to mineralogical nomenclature. The names of some of the most common and better known minerals have come from other European sources. Many of these names have obscure and uncertain origins, and frequently long etymologies. A few other terms have a modern origin, having been formed by adding the suffix -ite to common words of various languages.

ENGLISH

Many common English names of minerals, including several native elements, have Old English (Anglo-Saxon) origins. These originated with the early Teutonic settlers of Britain (from about A.D. 449 to A.D. 1100). The following terms can be traced to Old English: gold, ice, iron, lead, lime, silver, spar, tin, and water. Because of their common origin, some of these words are strikingly similar to the German words for the same substances: English *gold*, German *Gold;* English *silver*, German *Silber;* English *spar*, German *Spat;* English *tin*, German *Zinn;* and English *water*, German *Wasser.*

Although the term spar is now seldom used in scientific mineral names, it frequently was employed in the common names for some minerals. The word is derived from the Old English *spærstan, spar stone,* a name for gypsum. Later the term was used for many minerals which are vitreous, crystalline, and easily cleavable: for example, calcspar (calcite), feldspar, fluorspar (fluorite), heavy spar (barite), Iceland spar (calcite), pearl spar (dolomite), and tabular spar (wollastonite).

The names of several minerals can be understood by their similarities to modern English words. Although in some of these examples the names are not derived directly from English, the English words and the mineral names have parallel origins from such other languages as Latin, French, and Spanish. Names for colors include azurite (azure blue), carminite (carmine red), greenalite (green), and olivine (olive green). Smaltite received its name from the fact that it is the source of the pigment smalt, a deep blue cobalt oxide. Others are named for things they resemble, like gummite (gumlike), bronzite (bronze luster), and serpentine (surface often resembles the skin of a serpent).

GERMAN

Only a casual review of old mineral synonyms will show the tremendous influence the Germans had on the science of mineralogy. One will see many mineral terms ending with suffixes like *-erz, -glanz, -glimmer, -kalk, -kies, -salz, -spat,* and *-stein.* Although some of these terms are still used by German mineralogists, most of them have not been adopted by the international community. However, a few German names are very important in mineralogical literature.

hornblende, from an old German word for any dark prismatic mineral occurring in ores but containing no recoverable metal; from *horn,* for horn, possibly referring to the shape or color of horn, and *blenden,* meaning *to blind* or *deceive.*

nickel, from German *Nickel,* a name for Satan or "Old Nick," and from the mineral kupfernickel, *Old Nick's copper,* a substance equivalent to modern nickeline (NiAs) which was erroneously thought to contain copper, but would not give up any despite intensive smelting.

quartz, perhaps from the Saxon word *querkluftertz,* meaning *cross-vein-ore,* which could have become *querertz* and then quartz; this origin is supported by the old Cornish name for quartz, "cross-course-spar."

schorl, from German *Schörl,* perhaps derived from the locality Schörlau (meaning Schörl village) in Germany; some believe the name is a miner's term of unknown origin, and that the village got its name from the mineral; perhaps from the Old German word *Schor,* meaning *refuse* or *impurities;* originally applied to a large number of minerals which formed crystals of similar columnar shape.

wolframite, perhaps from *wolf,* meaning *wolf,* and *rahm,* meaning *froth* or *cream,* alluding to the objectionable scum formed during the smelting of tin ores containing tungsten as an impurity; or from *Wolfrig,* or some similar term, used by early Saxon miners in allusion to the action of the mineral, when present in tin ores, to decrease (to eat away or devour) the yield of tin during concentration or smelting.

zinc, from German *Zink,* a word of obscure origin.

RUSSIAN AND OTHER LANGUAGES OF THE SOVIET UNION

In spite of the fact the Russians have described many new minerals, especially in recent decades, not many Russian words have been incorporated in mineral names. Names introduced by Russian mineralogists are usually derived from the names of people, places, or chemical compositions. However, it is worth noting that some names have come from the languages of the Soviet Union.

loparite, from the Russian name for the Lapp inhabitants of the Kola Peninsula.

muscovite, from Russian *Moskva,* and then *muscovy glass,* a name used when the mineral was first described from Old Russia (Muscovy).

nordite is from Russian *nord, north,* because of its northern origin in the Lovozero tundras.

penkvilksite, from the Lapp words *penk*, meaning *curly*, and *vilkis*, meaning *white*, in allusion to its appearance.

sakhaite is from a name for Siberia in the Yakutian language.

tyuyamunite is from the name of a hill (Tyuya Muyun, meaning camel's humps) in Ferghana, Turkestan.

ROMANCE LANGUAGES

Although the etymologies of most of these terms go back to Latin and Greek, the names are generally associated with the languages of France, Spain, and Italy.

Several French terms have been used in mineral names. Celadonite is from *celadon*, meaning *sea green;* citrine, from *citron* (= *lemon*), was named for its yellow color; emery (*emeri*) has a long history going back to Greece; nacrite comes from *nacre*, meaning *mother-of-pearl;* turquoise implies *Turkish stone*.

The name jade comes from Spanish. It is derived from *piedra de yjada*, meaning *stone of the side*, so called because it was supposed to cure side pains or colic. Platinum is derived from *plata*, the Spanish word for silver.

Aventurine is from the Italian word, *avventura*, meaning *chance*, from the accidental discovery of a spangled glass.

ARABIC

The names of some well-known minerals were adopted by European mineralogists from Arabic words. In some cases these terms have long etymologies traceable to the early Persians. Realgar comes from *Rahj al ghar, powder of the mine;* talc comes from *talq*, meaning talc. Marcasite and trona apparently also have Arabic origins. The mineral serendibite is derived from Serendib, an old Arabic name for Ceylon. Arabic mineral names which were apparently borrowed from Persian include borax, lazulite, and zircon.

SANSKRIT AND OTHER LANGUAGES OF INDIA

Some ancient and classical mineral names have origins which have been traced to the languages of India, especially to Sanskrit. In most of these cases the names were adopted in turn by the Greeks and then by the Romans. Some of these names include opal (meaning precious stone), sapphire, tincal (meaning borax; the name of the mineral tincalconite is derived from this), and corundum (meaning ruby). More recent mineral names associated with Sanskrit are sinhalite (from Sinhala, the Sanskrit name for Ceylon), and cheralite (from Chera, the ancient Dravidian kingdom which was located in modern Travancore, India).

MISCELLANEOUS LANGUAGES

Many other languages have contributed to the nomenclature of minerals.

cymrite, from Welsh *Cymru,* a name for Wales.

feldspar, from the Swedish names for *field* and *spar,* in reference to the spar in tilled fields overlying granite.

kaolinite, from Chinese *kauling* (= *high ridge*), from the name of a hill near Jauchu Fa where the material was obtained.

makatite is from the Masai (Kenya, Africa) word *emakat* (= *soda*), in allusion to the high sodium content of the mineral.

stenhuggarite, from Swedish *stenhuggar* (= *stonemason*), in honor of Brian H. Mason, who has contributed much to the study of minerals of Langban, Sweden.

tacharanite, from Gaelic *tacharan* (= *changeling*), because on standing in air the mineral breaks down to form other minerals.

tourmaline, from *turamali,* a Singhalese word applied to zircon by jewelers of Ceylon.

xocomecatlite, from a Nahua (Mexican Indian) word for grapes, in allusion to the clusters of green spherules.

9
Prefixes Used in Mineral Names

Numerous prefixes, derived from Greek, Latin, and occasionally other languages, are found in mineral names. These are used either to modify previously existing mineral names or to form entirely new words. Generally the prefixes relate to colors, crystallographic properties, miscellaneous features, or chemical compositions.

PREFIXES RELATING TO PHYSICAL PROPERTIES

Commonly used color prefixes derived from Greek include the following.

chlor-, chloro-, green or light green.

chryso-, gold (generally used for yellow colors).

cyano-, kyan-, blue or dark blue.

glauco-, blue or bluish gray.

hema-, blood red.

leuco-, white or lack of color.

melano-, black.

rhodo-, rose.

xantho-, yellow.

Greek prefixes related to crystallographic properties are listed below.

clino-, slope or incline, usually in reference to monoclinic crystal system.

hemi-, half, used in reference to hemimorphic symmetry.

hexa-, six, in reference to hexagonal crystal system or water content.

iso-, equal, usually in reference to isometric crystal system.

ortho-, straight or right, usually in reference to orthorhombic crystal system.

tetra-, four, usually in reference to tetragonal crystal system.

tri-, three, often in reference to threefold symmetry.

MISCELLANEOUS PREFIXES

Several Greek prefixes are used to connote miscellaneous features of various types. Some of the more important ones are listed here.

a-, an-, not or without.

crypto-, concealed.

epi-, near or beside.

eu-, good or well.

hetero-, other or different.

hyalo-, glass.

lepido-, scale or flake.

meta-, along with, with, after, over, or changed in form.

para-, beside, nearby, beyond, or along with.

poly-, many or much.

pseudo-, false.

pyro-, fire or heat.

Of these prefixes, *meta-* and *para-* have been used very frequently to form names of new minerals which are closely related to older minerals. The new mineral may either have a slightly different chemical composition, for example with less water of crystallization (common use of *meta-*), or it might be a dimorphous structure related to the earlier mineral, or it might be associated with the older mineral in nature. See Chapter 4 for some examples of these.

CHEMICAL PREFIXES

By far the largest number of prefixes used in mineral names indicate something about the chemical composition, and have been used to form the names of numerous species. These prefixes have various origins, depending upon the origin of the chemical element name expressed; most, however, are from Greek or Latin.

alumino-, alumo-, aluminum, Al.

ammonio-, ammonium, NH_4^+

argento-, silver, Ag.

arseno-, arsenic, As.

auri-, auro-, gold, Au.

baryto-, barium, Ba; literally means *heavy*.

bismuto-, bismuth, Bi.

calcio-, calcium, Ca.

carbo-, carbonate, CO_3^{2-}.

chalco-, copper, Cu.

chlor-, chloro-, chlorine, Cl; also used for green color.

cupro-, copper, Cu.

ferri-, iron, Fe; usually ferric (III, 3+) valence.

ferro-, iron, Fe; usually ferrous (II, 2+) valence.

fluo-, fluor-, fluorine, F.

hydro-, water, H_2O.

iod-, *iodo-*, iodine, I.

kali-, *kalio-*, potassium, K.

lithio-, lithium, Li; literally means *stone*.

magnesio-, magnesium, Mg.

mangano-, manganese, Mn.

natro-, sodium, Na.

nitro-, nitrogen, N.

phospho-, phosphorus, P.

plumbo-, lead, Pb.

sidero-, iron, Fe.

stanno-, tin, Sn.

stib-, *stibio-*, antimony, Sb.

thor-, *thoro-*, thorium, Th.

urano-, uranium, U.

yttro-, yttrium, Y.

zinc-, *zinco-*, zinc, Zn.

zirco-, zirconium, Zr.

An additional chemical prefix, *molybdo-*, which is derived from the Greek word for lead, is ambiguous. (The element molybdenum was named for its resemblance to lead.) Some minerals using the *molybdo-* prefix contain lead, while others contain molybdenum.

Brief discussions on the use of chemical prefixes are given in Chapter 4 and Chapter 5.

10
Suffixes Used in Mineral Names

An examination of mineral names will show that the vast majority of them end in the suffix *-ite*. However, one will see a lack of uniformity in terminations, with names ending in *-lite, -ine, -an, -ar, -clase, -ide, -oid,* etc. For the sake of uniformity, it probably would be much better if all minerals ended in *-ite*. In fact, in the last two centuries various attempts have been made to convert old mineral names like galena to galenite, gypsum to gypsite, and spodumene to spodumenite, but generally these moves have not been successful. Chester [1] has reviewed some of this history.

COMMON SUFFIXES

The suffix *-ite* is derived from Greek and denotes *of the nature of* or *similar to*. It was used by the Greeks and Romans in mineral and rock terms, signifying a quality, constituent, use, or locality of the stone. Some ancient examples of this are *alabastrites* (stone out of which the vase called *alabastron* was made), *chloritis* (from its green color), *haimatitis* (from its blood-red color), and *syenitis* (rock from Syene, in Egypt). In some of the older literature one will find the ending *-yte*. Dana [2] proposed that the *-yte* spelling of *-ite* be used for rock names (for example, dioryte, tonalyte) to separate them from mineral names. This difference in spelling never caught on.

Another important suffix is -*lite*, derived from the Greek word for stone, *lithos*. This ending on mineral names also comes from the ancients, being found in some early Greek words. The use of this ending has not become popular except where it develops a more euphonious sounding word or where its meaning is beneficial, e.g., cryolite (meaning *ice* or *frost stone*). Also it should be emphasized that the termination -*ite* is not derived from -*lite*, and they are not related.

LESS COMMON SUFFIXES

The ending -*ine* is found in many names, for example, almandine, nepheline, olivine, sanidine, sapphirine, steenstrupine, and tourmaline. The ending is derived from Latin and Greek and means *pertaining to* or *like*. Basically it is close to -*ite* in significance. In some cases it has been changed to -*yne*, as in davyne (from H. Davy), hauyne (from R. J. Haüy), and levyne (from A. Lévy), because of the spelling of the name from which the mineral name was derived.

Another suffix derived from Latin and Greek is -*oid*, which generally means *like*. It appears in the terms chloritoid, feldspathoid, and pyroxenoid.

The ending -*an*, derived from French and Latin, commonly means *belonging to* or *pertaining to*. It is used to form numerous mineral names including celsian, lavendulan, nosean, and retzian.

The common Latin termination -*um* is traditionally used for the names of the chemical elements. Native elements indium, iridium, platinum, selenium, and tellurium are examples of this. Other mineral names with this Latin ending include electrum and minium.

Numerous suffixes, derived from Greek or Latin terms, have specialized applications. The term -*clase* (Greek, = *fracture* or *break*), forms words like clinoclase, orthoclase, plagioclase, and rhomboclase. A few names end with -*crase* (Greek, = *mixture*), for example, idocrase and polycrase. Microcline is from Greek -*cline* (= *slant*). Several names end with -*melane* (Greek, = *black*): cryptomelane, lepidomelane, and psilomelane. Others end with -*phane* (Greek, = *to appear*), for example, hedyphane, hyalophane, and uranophane. The suffix -*gen,* from Greek and Latin meaning *to be born* or *derived,* is used to form alunogen and botryogen. The ending -*ide,* used to form adjectives from Latin verbs and usually meaning *having the quality of,* is used to form names like auricupride.

NAMES WITHOUT SUFFIXES

R. J. Haüy (1743–1822), the famous French crystallographer and mineralogist, introduced many mineral names which, although they were often derived from Greek terms, did not usually end with the common suffixes. Dana [2] was critical of Haüy's departure from the common use of the -*ite* suffix; he wrote: "Forgetting that the unity of law which he [Haüy] had found in nature should be a feature of scientific language, he gave to his names the following terminations. . . ." Following this is a list of various inconsistent mineral names invented by Haüy, including terms such as amphibole, anatase, diallage, epidote, harmotome, pleonaste, pyroxene, and sphene.

Many other mineral names, especially those of ancient origin and derived from various languages, do not have consistent endings. Such names include beryl, borax, cinnabar, corundum, litharge, opal, quartz, realgar, talc, and many others. Likewise mineral names derived from Old English (Anglo-Saxon) have inconsistent endings, for example, gold, ice, iron, lead, lime, spar, and water.

REFERENCES

1. Chester, Albert H. *A Dictionary of the Names of Minerals.* New York: John Wiley & Sons, 1896.
2. Dana, James D. *System of Mineralogy,* Fifth Ed. New York: John Wiley & Son, 1868, p. xxxi.

11

The Use of Symbols with Mineral Names

In order to avoid a proliferation of new mineral names, especially in cases where there are only slight differences between minerals, symbols have been utilized. Symbols are employed to show both crystal structural variations and chemical variations.

POLYMORPHISM

Some chemical substances crystallize in two or more different structures. For example, native sulfur may be either orthorhombic or monoclinic. This phenomenon, very common in minerals, is generally called polymorphism (literally, *many forms*). Sometimes the specific terms dimorphism (*two forms*) or trimorphism (*three forms*) are used if they apply. Most polymorphic minerals are given specific names; for example, the four forms of carbon (C) are named graphite, diamond, lonsdaleite, and chaoite, and the three forms of calcium carbonate ($CaCO_3$) are named calcite, aragonite, and vaterite. However, another method for distinguishing between substances having more than one structure is by using Greek letters. These are usually prefixed to a mineral name. Common examples include the following: α-cristobalite, β-cristobalite; β-fergusonite; α-quartz, β-quartz; β-roselite; α-sulfur, β-sulfur, γ-sulfur. In some cases numerical subscripts have been used with the Greek letters to indicate additional structural differences, for example, β_1-tridymite and β_2-tridymite. Because the use of Greek letters is rather cumbersome, in recent years the names have been somewhat simplified by mineralogists, either by spelling the Greek letter, as in beta-fergusonite and beta-roselite, or by using the terms low, middle, and high, in reference to temperature stability ranges. Frondel,[1] for example, has used low-tridymite (for α-tridymite), middle-tridymite (for β_1-tridymite), and high-tridymite (for β_2-tridymite). In some cases a new name has been introduced, for example, rosickyite is a synonym for γ-sulfur.

POLYTYPISM

Another kind of symbol has been used to designate the various structural polytypes of certain minerals with layered structures. These minerals often occur in several modifications which differ only in the manner in which identical unit layers of structure are stacked upon each other. Varying stacking arrangements of the basic structural layer give rise to different unit cell dimensions (especially perpendicular to the stacked planes), and may give rise to different space groups and occasionally even different crystal systems. Usually, the different polytypes of a substance have nearly identical physical and chemical properties and can only be identified by special techniques like X-ray diffraction. The various polytypes of a given substance are generally represented by a symbol which usually designates two aspects of the structure: (1) the number of unit layers stacked upon each other to form the crystallographic unit cell perpendicular to the stacking, and (2) the crystal system or symmetry of the stacking arrangement. Common wurtzite, for example, has a hexagonal stacking involving two structural layers and is designated 2H. Other less common polytypes are designated as wurtzite-4H, wurtzite-6H, and wurtzite-15R. Other examples of polytypic minerals are: biotite-1M, biotite-2M; cronstedtite-1H, cronstedtite-$2H_a$, cronstedtite-$2H_b$, cronstedtite-$2H_c$, cronstedtite-6R; högbomite-5H, högbomite-18R; molybdenite-2H, molybdenite-3R; muscovite-1M, muscovite-$2M_1$, muscovite-$2M_2$, muscovite-$3T$. The letter symbols which have been used are C (cubic), H (hexagonal), T (trigonal), R (rhombohedral), O or OR (orthorhombic), M (monoclinic), and Tc or A (triclinic). Alphabetical or numerical subscripts are used with these capital letters for polytypes which consist of equivalent numbers of layers per unit cell, but in which the layers have different stacking arrangements, as is true of cronstedtite and muscovite mentioned above.

Another system for designating polytypes has been recommended recently.[2] This is based upon the concepts just described, but includes additional information. For example, molybdenite-$2H$ becomes molybdenite-$Haa2c$; molybdenite-$3R$ becomes molybdenite-$Raa3c$. The three lower-case letters, accompanied by numbers when necessary, following the symmetry symbol capital letter, indicate the periodicities of the three unit-cell edges (in the order a, b, c, or in hexagonal and rhombohedral, a, a, c) of the compound relative to those of the smallest parental subcell of the polytypic system. In the example given the first polytype is hexagonal and has a c dimension of two layers; the second structure is rhombohedral, it has the same two a dimensions as the first structure, but its c is three layers thick. In another example, similar symbols show both relationships and differences between the various pyrrhotite polytypes where the structures differ in orientation, crystal system, unit-cell periodicities, etc.: pyrrhotite-$Hbb2c$ (troilite); pyrrhotite-$H2a2a6c$ (old $6C$ type *); pyrrhotite-$OR2a2a11c$ (old $11C$ type); pyrrhotite-$H2a2a5c$ (old $5C$ type); pyrrhotite-$M2b2a4c$ (old $4C$ type). For additional examples and a more detailed discussion of this nomenclature see the summary by Bailey.[2]

CHEMICAL DIFFERENCES

In recent years the names of some of the rare-earth minerals have been modified by chemical element symbols, in order to indicate which rare earth predominates within the mineral. This is illustrated by allanite-(Ce), allanite-(La), and allanite-(Y). The name,

* In the old designation for pyrrhotite polytypes C refers to the c-axis rather than *cubic*.

without the suffixed symbol, implies a mineral of the allanite group, and that the proportion of rare-earth elements in the specimen has not been determined. When the rare-earth-element distribution has been determined, the chemical symbol for the predominant rare-earth element is appended, and the name is converted to one for a species. This method of nomenclature, devised by Levinson,[3] has been applied to many rare-earth mineral names, including the following: aeschynite-(Ce), aeschynite-(Y); monazite-(Ce), monazite-(La); britholite-(Ce), britholite-(Y).

Suffixed chemical symbols also have been used for designating members of isomorphous series which do not contain rare earths.[4] This can be illustrated by the name osumilite-(K,Mg). The mineral is closely related to the other osumilites, but in this case K exceeds Na, and Mg exceeds Fe. In other words it is close to the K,Mg end-member of the osumilite series. Unlike the rare-earth minerals, these names have not received general acceptance.

Povarennykh,[5] in an attempt to formulate rules for composing rational names for minerals, has suggested an extreme use of symbols in mineral names. He has advocated the introduction of Arabic numerals within the mineral name, especially where two or more minerals have similar qualitative compositions but differ quantitatively. For example, boulangerite ($Pb_5Sb_4S_{11}$) would become plum5stib4sulite, sterryite ($Pb_{12}Sb_{10}S_{27}$), containing the same elements, would become plum12stib10sulite, and playfairite ($Pb_{16}Sb_{18}S_{43}$) would become plum16stib18sulite. Further examples include owyheeite ($Ag_2Pb_5Sb_6S_{15}$) which would become ar2-plum5stib6sulite and nakaseite ($Ag_3CuPb_4Sb_{12}S_{24}$) which would become ar3cuplum4stib12sulite. Although these names are given in Povarennykh's book,[5] along with the standard names, they have not been adopted by the scientific community.

REFERENCES

1. Frondel, C. *System of Mineralogy of . . . Dana,* Seventh Ed., Volume III. New York: John Wiley & Sons, 1962.
2. Bailey, S. W. "Report of the I.M.A.–I.U.Cr. Joint Committee on Nomenclature." *American Mineralogist* 62: 411–415 (1977).

3. Levinson, A. A. "A system of nomenclature for rare-earth minerals." *American Mineralogist* **51:** 152–158 (1966).
4. Chinner, G. A., and Dixon, P. D. "Irish osumilite." *Mineralogical Magazine* **39:** 189–192 (1973).
5. Povarennykh, A. S. *Crystal Chemical Classification of Minerals,* Volumes I–II. New York: Plenum Press, 1972.

12
Rules Regarding the Formulation of Mineral Names

Deliberations on the validity of mineral names generally fall into two categories, those concerning old names in the literature and those concerning new names proposed for newly discovered minerals. Each of these aspects will be considered separately.

OLD NAMES

Often in the early days several names were published for the same mineral, because of difficulties in communication and lack of means for determining precise chemical and crystallographic data. The *law of priority*, which states that the oldest name should be retained, is applied in most of these cases. Dana,[1] however, has proposed some exceptions to this law, indicating when the law of priority may be set aside. The older name should not be used when it is identical to the accepted name of another mineral of earlier date. The name should not be used when it implies a property which is false; for example, the prior name melanochroite (meaning *black color*) was replaced by the newer name phoenicochroite (meaning *deep red color*) because the mineral is characteristically red. An older name should not have precedence if it was put forth without an adequate description. Likewise, if the original description of the mineral with the older name is so incorrect that a recognition of the mineral by means of it is impossible, the older name need not have priority. However, if a well-known but poorly described older mineral is redescribed correctly, the new describer should not change the old name. If the older name is based on an uncharacteristic variety of the species, the name should not be used; thus sagenite was properly set aside for rutile.

Other exceptions to the law of priority should also be considered. In some cases the status or meaning of conflicting names has been changed in such a way so they both can be retained. For example, the earlier name augite was assigned to the mineral species while the later name pyroxene was retained as a group name. We see then that the law of priority need not apply to some names which are not used for species, for example, amphibole, chlorite, garnet, mica, and pyroxene. Names derived through ignorance or carelessness should not be allowed to perpetuate blunders under any law of priority. Dana [1] was especially careful to point out problems which arise when names derived from Greek terms are put together badly; or which arise from names derived by combinations of Greek and Latin, or of common words from other languages. Also, it has been suggested that if an older name is lost sight of and has found no one to assert its claim for a period of more than fifty years, it should be dropped, especially if the later name adopted for the species has become intimately incorporated within the terminology of the science. In updating Dana's discussion, Palache, Berman, and Frondel [2] indicated that they preferred names after persons when a choice had to be made between a number of names for the same mineral.

Once an older name has been set aside it is not only considered obsolete, it is considered to be completely extinct and should never be used again. This is the only reasonable way to avoid confusion, for if the name were later applied to a different species it could well give rise to ambiguity. Consequently, many suitable names have fallen into disuse; and this is especially unfortunate where the names of important and famous mineralogists and other persons can never be used again. In recent years, as discussed previously, this problem has been circumvented by using first or middle names, or by modifying the family name; for example, although genthite has become obsolete (it is the same as garnierite), the name Genth is still used in the mineral genthelvite (Genth + helvite).

Although some mineral names which are synonyms have been retained as variety names—for example, cleavelandite (variety of albite), emerald (variety of beryl), kunzite (variety of spodumene), and ruby (variety of corundum)—the number of such varietal terms, especially those based on chemical differences, has been reduced by the use of chemical adjectival modifiers. Some examples of this trend include mercurian tetrahedrite for schwatzite; arsenian vanadinite for endlichite; chromian muscovite for fuchsite; manganoan fayalite for knebelite; etc. Here the adjectival modifier is determined by the constituent next in abundance after the principal constituents of the species. Many older variety names have been rendered obsolete in this way. The adjectival endings for the various chemical modifiers were proposed by Schaller [3] and are summarized by Palache, Berman, and Frondel.[2] The modifying adjectives also show the valence when more than one such state exists, as with iron (ferroan for ferrous iron, ferrian for ferric iron).

NEW NAMES

The names of new minerals also must follow some definite rules. First it must be established that the mineral is a new species, and not simply a variety of an older mineral.[4,5] The name for the new mineral must itself be completely new. It cannot be a resurrected old name, nor can it be a name similar to an old name, whether obsolete or current. One of the names in the following pairs probably would not have been allowed if it had been proposed at the present time because of possible ambiguities: berthierine and berthierite; danaite and danalite; daubreeite and daubreelite; fersmanite and fersmite.

Several other guidelines, summarized by Dana [1] and Palache, Berman, and Frondel,[2] have been formulated. As far as practicable, names should terminate uniformly in *-ite*, or in some cases in *-lite*. In forming such names from the Greek or Latin the suffix is added to the genitive form after dropping the vowel or vowels of the last syllable, and any following letters. Greek is the preferred language for such names. In the formation of the names of minerals, the addition of the -*ite* to *proper names* in modern languages (names of places or persons), or names of characteristic chemical constituents, is allowable; but attaching -*ite* to *common words* in modern languages is usually considered objectionable. It is also considered poor practice to form mineral names combining Greek and Latin, or Greek or Latin and a modern language.

Also, double word names should not be used. The old names gay-lussite and sal ammoniac, for example, have become gaylussite and salammoniac, and all new names should follow a similar pattern. Mineral names derived from persons named *Mc-* and *Mac-* have caused considerable debate.[6] Most mineralogists would like to use *mac-* uniformly for mineral names, like macconnellite. However, it has been decided to use the original form of the personal name, so the term actually used is mcconnellite (for R. B. McConnell). In contrast, macdonaldite (for G. A. Macdonald) is retained.

When a substance is shown to consist of several structural polytypes, these are not given separate names, but are designated by appropriate suffixed symbols (wurtzite-$2H$ and wurtzite-$15R$). In like manner, the terminology of some of the rare-earth minerals has been simplified by using chemical element symbols as suffixes; for example, instead of using priorite, the name aeschynite-(Y) now applies. The use of symbols such as these with mineral names is discussed in Chapter 11.

TRANSLATION OF NAMES INTO ENGLISH

Because of language differences there are some problems in converting mineral names from other modern languages into English. Although these rules have not been adhered to rigidly, some of the guidelines which have been suggested are outlined here.

- The diphthongs *ae* and *oe* can be simplified to *e*.
- The German umlaut can be replaced by an *e* after the vowel.
- The Swedish *å* can be reduced to *a*, while the Norwegian *å* can be replaced by *aa*.
- Accents and various diacritical marks generally are not retained.

- When letters in a name are brought together in a manner not customary in English, a change can and should be made.
- For Russian names the British Standard transliteration system is commonly used, although some prefer the rules of the Library of Congress.[7]
- There are problems in translating Chinese names. The P'in-Yin Romanization System is widely used in the People's Republic of China; however, the Wade-Giles Romanization System, more widely used in the western scientific community, has the advantage of facilitating a proper pronunciation.[8]

Unfortunately, guidelines like these have not been agreed upon completely by the English-speaking community. The British, for example, have not accepted the simplification of diphthongs, still using haematite, for example. Within the United States the treatment of the umlaut is not uniform; in the literature we find both roesslerite and rösslerite, both kroehnkite and kröhnkite.

So far the Commission on New Minerals and Mineral Names, of the International Mineralogical Association, has not formulated a set of rules about many of the matters discussed in this chapter.

REFERENCES

1. Dana, Edward S. *System of Mineralogy of . . . Dana,* Sixth Ed. New York: John Wiley & Sons, 1892.
2. Palache, C., Berman, H., and Frondel, C. *System of Mineralogy of . . . Dana,* Seventh Ed., Volume I. New York: John Wiley & Sons, 1944.
3. Schaller, W. T. "Adjectival ending of chemical elements used as modifiers to mineral names." *American Mineralogist* **15:** 566–574 (1930).
4. Hey, M. H., Guillemin, C., Permingeat, F., and de Roever, J. P. "Sur la nomenclature minéralogique." *Bulletin de la société française de minéralogie et de cristallographie* **84:** 96–97 (1961).
5. Fleischer, Michael. "Procedure of the International Mineralogical Association Commission on New Minerals and Mineral Names." *American Mineralogist* **55:** 1016–1017 (1970).
6. Fleischer, Michael. "What's in a name; Mac vs. Mc in mineral names." *Mineralogical Record* **3:** 235–236 (1972).
7. Allen, Charles G. *A Manual of European Languages for Librarians.* London and New York: Bowker, 1975.
8. Newnham, Richard. *About Chinese.* Middlesex: Penguin Books, Ltd., 1971.

Part II

Mineral Names:
An Alphabetical List

A

Abelsonite, for Philip Hauge Abelson (1913–), American organic geochemist, president of the Carnegie Institution, Washington, D.C., and editor of *Science.*

Abernathyite, for Jess Abernathy, operator of the Fuemrol mine, Emery County, Utah, who first found the mineral.

Acanthite, from Greek for *thorn,* in allusion to the shape of the crystals.

Acetamide, from Latin *acetum* (= *vinegar*), and *amide,* a contraction of *am*monia and *ide.*

Achroite (var. of elbaite), from Greek for *not* and *color,* in allusion to its colorless nature.

Acmite, from Greek for *point,* in reference to its pointed crystal habit.

Actinolite, from Greek for *ray,* in allusion to its frequent occurrence in bundles of radiating needles.

Adamite, for Gilbert Joseph Adam (1795–1881), French mineralogist, who supplied the first specimens for examination.

Adelite, from Greek word meaning *obscure* or *indistinct,* for its lack of transparency.

Adularia (var. of orthoclase), for Adular Mountains in the Alps, erroneously supposed to be the range containing St. Gotthard, its locality.

Aegirine (syn. for acmite), for Ægir, the Scandinavian god of the sea, the mineral being reported first in Norway.

Aenigmatite, from Greek word for *riddle,* apparently alluding to its initial problematical chemical nature.

Aerugite, from Latin *aerugo* =*rust of copper,* alluding to its green color, although it does not contain copper.

Aeschynite, from Greek for *shame,* in allusion to the inability of chemical science, at the time of its discovery, to separate some of its constituents.

Afghanite, locality in Afghanistan, the country in which the mineral was first found.

Afwillite, for Alpheus Fuller Williams (1874–?), of the DeBeers Consolidated Mines, Kimberley, South Africa.

Agardite, for Jules Agard, French geologist, Orléans.

Agate (var. of quartz), locality in the river Achates, perhaps the modern Carabi or Cannitello, in Sicily.

Agrellite, for Stuart O. Agrell, English mineralogist, Cambridge University.

Agrinierite, for Henri Agrinier (1928–1971), engineer in the mineralogy laboratory of the Commission Energie Atomique, France.

Aguilarite, for P. Aguilar, superintendent of the San Carlos mine, Guanajuato, Mexico.

Ahlfeldite, for Friedrich Ahlfeld (1892–　), German mining engineer and mineralogist, who studied the minerals of Bolivia.

Aikinite, probably for Arthur Aikin (1773–1854), a founder of the Geological Society of London.

Ajoite, locality at Ajo, Pima County, Arizona.

Akaganeite, locality at Akagane mine, Iwate Prefecture, Japan.

Akatoreite, locality at Akatore Creek, Eastern Otago, New Zealand.

Akdalaite, from the Kazakh name of the locality in the Karagandin region of Kazakhstan, Soviet Union.

Akermanite, for Anders Richard Åkerman (1837–1922), Swedish metallurgist.

Akrochordite, from Greek word for *wart,* in reference to the appearance of the aggregates of the mineral.

Aksaite, locality at Aksai, Kazakhstan, Soviet Union.

Aktashite, locality at Aktash mercury deposit, Gornyi Altai, Soviet Union.

Alabandite, locality at Alabanda in Caria, Asia Minor.

Alabaster (var. of gypsum), ancient name originally given to several materials from which the ointment vases called alabastra were made; perhaps from Alabastron, a town in Egypt.

Alamosite, locality at Alamos, Sonora, Mexico.

Albite, from Latin *albus* (= *white*), in allusion to the color.

Albrittonite, for Claude Carroll Albritton, Jr. (1913–　), American geologist, Southern Methodist University, Dallas.

Aldzhanite, locality not specified in the Soviet Union.

Alexandrite (var. of chrysoberyl), for Czar Alexander II (1818–1881) of Russia.

Algodonite, locality at silver mine of Algodones, near Coquimbo, Chile.

Aliettite, for Andrea Alietti (1923–　), Italian mineralogist, University of Modena.

Allactite, from Greek word meaning *to change,* in allusion to its strong pleochroism.

Allanite, for Thomas Allan (1777–1833), Scottish mineralogist, who first observed the mineral.

Allargentum, from Greek word for *another* and Latin *argentum* for *silver.*

Alleghanyite, locality in Alleghany County, North Carolina.

Alloclasite, from Greek words for *another* and *fracture,* because its cleavage was believed to differ from other minerals it resembles.

Allophane, from Greek words for *another* and *to appear,* in allusion to its change of appearance under the blowpipe.

Alluaudite, for its discoverer François Alluaud, of Limoges, France.

Almandine, locality at Alabanda, in Asia Minor, where garnets were cut and polished in ancient times.

Alstonite, locality near Alston Moor, Cumberland, England.

Altaite, locality in Altai Mountains, Siberia, Soviet Union.

Althausite, for Egon Althaus, professor, Univeristy of Karlsruhe, Germany.

Alum (mineral group name), from Latin *alumen,* its ancient name.

Aluminite, from composition, high in *alumina.*

Aluminocopiapite, from composition, like copiapite, but with predominant aluminum.

Alumohydrocalcite, from composition, a *hydr*ated carbonate of *calc*ium and *alum*inum.

Alumotungstite, from composition, *alum*inum and *tungst*en.

Alunite, from Latin *alumen* (= *alum*).

Alunogen, from Latin *alumen* (= *alum*) and Greek for *to spawn.*

Alvanite, from composition, *al*uminum and *van*adium.

Amakinite, for the Amakin Expedition, which prospected the diamond deposits of Yakutia, Soviet Union.

Amalgam (alloy of mercury and silver), from Greek for *emollient* or *poultice,* related to *malakos* (= *soft*).

Amarantite, from Greek *amaranthos* (= *amaranth*), alluding to its red color.

Amarillite, locality at Tierra Amarilla, Chile.

Amazonite (var. of microcline), locality at Amazon River, South America.

Amblygonite, from Greek for *blunt* and *angle,* in allusion to the cleavage angle.

Ameghinite, for Florentino Ameghino (1854–1911) and Carlos Ameghino (1865–1936), Argentine geologists.

Amesite, for James Ames, a mine owner.

Amethyst (var. of quartz), from Greek for *not* and *drunk* because it was supposed to have the power to remedy drunkenness.

Aminoffite, for Gregori Aminoff (1883–1947), Swedish mineralogist and crystallographer, Riksmuseum, Stockholm.

Ammonia alum, from its composition, the ammonium member of the alum group.

Ammonioborite, from its composition, *ammon*ium and *bor*ate.

Ammoniojarosite, from composition, like jarosite, but with predominant ammonium.

Amosite (var. of cummingtonite), from the initial letters of Asbestos Mines of South Africa (AMOS).

Amphibole (mineral group name), from Greek for *ambiguous,* in allusion to the great variety of composition and appearance shown by this group of minerals.

Analcime, from Greek for *weak,* in allusion to its weak electric property when heated or rubbed.

Anandite, for Ananda Kentish Coomaraswamy (1877–1947), the first director of the mineral survey of Ceylon.

Anapaite, locality in the mines near Anapa on the Taman Peninsula, Black Sea, Soviet Union.

Anatase, from Greek for *extension,* in allusion to the greater length of the common pyramid as compared to other similar tetragonal minerals.

Anauxite (var. of kaolinite), from Greek words meaning *not expanding,* because the mineral does not swell when heated with a blowpipe.

Ancylite, from Greek word for *curved,* in allusion to the rounded and distorted character of the crystals.

Andalusite, locality in Andalusia, a province of Spain.

Andersonite, for Charles Alfred Anderson (1902–   ), American geologist, U.S. Geological Survey.

Andesine, locality in the Andes Mountains, where it is the chief feldspar in the andesite lavas.

Andorite, for Andor von Semsey (1833–1923), Hungarian nobleman interested in minerals and meteorites.

Andradite, for J. B. d'Andrada e Silva (1763–1838), Brazilian mineralogist, who examined and described a variety of this mineral.

Andremeyerite, for André Marie Meyer (1890–   ), Belgian mining engineer, honorary director general of mining, who collected the original specimens in Zaire.

Andrewsite, for Thomas Andrews (1813–1885), Irish chemist.

Angelellite, for Victorio Angelelli, Argentine mineralogist.

Anglesite, locality on Island of Anglesey, Wales.

Anhydrite, from Greek for *without water,* in contrast to the more common calcium sulfate, gypsum, which contains much water.

Anilite, locality at Ani mine, Akita, Japan.

Ankerite, for Mathias Joseph Anker (1771–1843), Austrian mineralogist.

Annabergite, locality at Annaberg, Saxony.

Anorthite, from Greek for *not straight*, because of its triclinic symmetry.

Anorthoclase, from Greek for *not straight* and *fracture*, in allusion to its cleavage angle.

Antarcticite, locality on the continent of Antarctica.

Anthoinite, for Raymond Anthoine (1884–?), Belgian geologist.

Anthonyite, for John Williams Anthony (1920–   ), American geologist, University of Arizona.

Anthophyllite, from Neo-Latin *anthophyllum*, meaning *clove*, in allusion to the clove-brown color of the mineral.

Antigorite, locality at Antigorio in Italy.

Antimonpearceite, from composition, like pearceite, but with predominant antimony.

Antimony, from Medieval Latin *antimonium*, possibly of Arabic origin.

Antlerite, locality at Antler mine, Mohave County, Arizona.

Apatite (mineral group name), from Greek for *to deceive*, because the gem varieties were confused with other minerals.

Aphthitalite, from Greek for *unalterable* and for *salt*, because it is stable in air.

Apjohnite, for James Apjohn (1796–1886), Irish chemist and mineralogist, Trinity College, Dublin

Aplowite, for Albert Peter Low (1861–1942), Canadian geologist, director of the Geological Survey of Canada, Ottawa.

Apophyllite (mineral group name), from Greek for *away from* and *leaf*, because of its tendency to exfoliate when heated.

Aquamarine (var. of beryl), from Latin *aqua marina* (= *seawater*), in allusion to its color.

Aragonite, locality at Aragon, Spain, where the pseudohexagonal twinned crystals were first recognized.

Aramayoite, for Félix Avelino Aramayo (1846–1929), a mine director in Bolivia.

Arcanite, from the Medieval Latin alchemical name for the compound, *Arcanum duplicatum* (= *double secret*).

Archerite, for Michael Archer, Curator of Mammals, Queensland Museum, Australia.

Arcubisite, from composition, silver (*argentum*), copper (*cuprum*), bismuth, and *sulfur.

Ardealite, locality at Ardeal, the old Rumanian name for Transylvania where it was initially found.

Ardennite, locality near Ottrez, in the Ardennes, Belgium.

Arfvedsonite, for Johan A. Arfvedson (1792–1841), Swedish chemist.

Argentite, from Latin *argentum* (= *silver*), in allusion to its composition.

Argentojarosite, from composition, like jarosite, but with predominant silver (*argent*um).

Argentopyrite, from composition, a silver (*argen*tum) -bearing mineral similar to pyrite in appearance.

Argyrodite, from Greek for *silver-containing,* in allusion to its composition.

Aristarainite, for Lorenzo Francisco Aristarain, Universidad Nacional de la Plata, Argentina, who has contributed to borate mineralogy.

Armalcolite, for the astronauts responsible for its collection on the Moon: Neil Alden *Arm*strong (1930–   ), Edwin Eugene *Al*drin, Jr. (1930–   ), and Michael *Col*lins (1930–   ).

Armangite, from its composition, *ar*senic and *mang*anese.

Armenite, locality at Armen mine near Kongsberg, Norway.

Armstrongite, for Neil Alden Armstrong (1930–   ), American astronaut, first man on the Moon's surface (Apollo XI lunar mission, 1969); not a lunar mineral.

Arnimite, for the von Arnim family, owners of the Planitz coal works near Zwickau, Germany, the locality of the mineral.

Arrojadite, for Miguel Arrojado Ribeiro Lisboa (1872–1932), Brazilian geologist.

Arsenate-belovite, from composition, an arsenate, and for Nikolai Vasilevich Belov (1891–   ), Russian crystallographer.

Arsenic, from Greek word for *strong* or *masculine,* from the belief that metals were of different sexes, or because it was known to have potent properties.

Arseniosiderite, from composition, *ar*senic and iron (*sideros*).

Arsenobismite, from composition, *arsen*ic and *bism*uth.

Arsenoclasite, from composition, *arsen*ic, and Greek for *fracture,* in allusion to its excellent cleavage.

Arsenolamprite, from composition, *arsen*ic, and Greek for *brilliant,* in allusion to its bright metallic luster.

Arsenolite, from composition, *arsen*ic.

Arsenopalladinite, from composition, *arsen*ic and *pallad*ium.

Arsenopyrite, from composition, an *arsen*ic bearing mineral similar to pyrite in appearance; contraction of the older term arsenical pyrites.

Arsenosulvanite, from composition, like sulvanite, but with predominant arsenic.

Arsenpolybasite, from composition, like polybasite, but with predominant arsenic.

Arsenuranospathite, from composition, *arsen*ic and *uran*ium, and Greek word for *a broad blade,* in allusion to the crystal habit.

Arsenuranylite, from composition, *arsen*ic and *uranyl,* and its relationship to phosphuranylite.

Arthurite, for Arthur Edward Ian Montagu Russell (1878–1964), English mineralogist and collector of Reading, and Arthur William Gerald Kingsbury (1906–1968), English mineralogist, Oxford.

Artinite, for Ettore Artini (1866–1928), Italian mineralogist, University of Milan.

Asbecasite, from composition, *As, Be, Ca,* and *Si.*

Asbestos (general term for several asbestiform minerals), from Greek for *inextinguishable,* after the misconception that once ignited it could not be extinguished.

Ashcroftine, for Frederick Noel Ashcroft (1878–1949), mineral collector of London and donor of many zeolites to the British Museum.

Astrophyllite, from Greek for *star* and *leaf,* in allusion to the stellate aggregation and foliated micaceous structure of the mineral.

Atacamite, locality in the province of Atacama in Chile.

Atelestite, from Greek word for *incomplete,* presumably because its composition was unknown when it was first described.

Athabascaite, locality at Lake Athabasca in northern Saskatchewan and Alberta, Canada.

Atheneite, for the Greek goddess Pallas Athene, in reference to the palladium content of the mineral; element palladium has name derived from same goddess.

Atokite, locality at Atok platinum mine in the Bushveld Igneous Complex, South Africa.

Attakolite, from Greek for *salmon,* alluding to the pale red color.

Augelite, from Greek for *luster,* in allusion to its characteristic pearly luster on cleavage surfaces.

Augite, from Greek for *luster,* from the appearance of cleavage planes.

Aurichalcite, either from Greek for *mountain copper,* or from Latin for *gold* and Greek for *copper* (it neither contains gold, nor has a golden color).

Auricupride, from composition, gold (*aur*um) and copper (*cupr*um).

Aurorite, locality at Aurora mine, Treasure Hill, Hamilton, Nevada.

Aurostibite, from composition, gold (*aur*um) and antimony (*stib*ium).

Austinite, for Austin Flint Rogers (1877–1957), American mineralogist, Stanford University.

Autunite, locality at Autun, France.

Aventurine (var. of quartz or feldspar), from Italian *avventura* (= *chance*), from the accidental discovery of the synthetic material.

Avicennite, for Abu Ali Ibn Sina [Avicenna] (980–1037), Arab physician and scientist, who lived in Bukhara, Persia (now in the Soviet Union), where the mineral was discovered.

Avogadrite, for Amadeo Avogadro (1776–1856), physicist, University of Turin, Italy.

Axinite, from Greek word meaning *ax,* in allusion to the wedgelike shape of the crystals.

Azoproite, from the Russian title for the International Geological Association for the Study of Deep Zones of the Earth's Crust (AZOPRO), because it was found during the preparation of a guidebook for the Association's meeting at Baikal in 1969.

Azurite, from the Persian word *lazhward,* meaning blue, in allusion to its color.

B

Babefphite, from composition; contains *Ba, Be, F, P,* and *H.*

Babingtonite, for William Babington (1757–1833), Irish physician and mineralogist.

Baddeleyite, for Joseph Baddeley, who brought the original specimens from Ceylon.

Badenite, locality near Badeni-Ungureni, Muscel, Rumania.

Bafertisite, from composition, *Ba,* iron (*ferrum*), *Ti,* and *Si.*

Bahianite, locality at the Serra de Mangabeira, state of Bahia, Brazil.

Bakerite, for R. C. Baker, of Nutfield, Surrey, England, who discovered the mineral in California.

Balipholite, from the Chinese name meaning *fibrous, barium-lithium* mineral; the first four letters refer to the chemical elements barium (*Ba*) and lithium (*Li*).

Balkanite, locality in Balkan Peninsula, in southern Europe.

Bambollaite, locality at Mina La Bambolla, near Moctezuma, Sonora, Mexico.

Banalsite, from composition, containing *Ba, Na, Al,* and *Si.*

Bandylite, for Mark Chance Bandy (1900–1963), American mining engineer, who collected the mineral.

Bannisterite, for Frederick Allen Bannister (1901–   ), keeper of minerals, British Museum of Natural History, London.

Baotite, locality at Paotow (Russian, *Baotoy*), Inner Mongolia.

Bararite, locality at Barari, Bengal, India.

Baratovite, for Rauf Baratovich Baratov, Soviet petrographer, Tadzhikistan.

Barbertonite, locality in Barberton district of Transvaal, Union of South Africa.

Barbosalite, for Aliuzio Licinio de Mirande Barbosa, Brazilian geologist, Escola de Minas, Minas Gerais.

Bariandite, for Pierre Bariand, French mineralogist, curator, University of Paris.

Baricite, for Ljudevit Barić, Yugoslavian mineralogist, professor, University of Zagreb.

Barite, from Greek word meaning *heavy*, in allusion to its relatively high specific gravity.

Barkevikite, locality at Barkevik, Langesundfjord, Norway.

Barnesite, for William Howard Barnes (1903–), National Research Council of Canada, who studied crystal structures of vanadium minerals.

Barrerite, for Richard Maling Barrer (1910–), British chemist, Imperial College, London, who has contributed to the chemistry of molecular sieves.

Barringerite, for Daniel Moreau Barringer (1860–1929), American mining engineer, an early defender of the meteorite origin of the Barringer or Meteor crater of Arizona.

Barringtonite, locality at Barrington Tops, New South Wales, Australia.

Barylite, from Greek for *heavy*, in allusion to its high specific gravity.

Barysilite, from Greek word for *heavy*, and from composition, containing *sili*con.

Barytocalcite, from composition; contains *bar*ium and *calc*ium.

Barytolamprophyllite, from composition, like lamprophyllite, but with predominant barium.

Basaluminite, from composition, a *bas*ic sulfate of *alumin*um; similar to aluminite.

Bassanite, for Francesco Bassani (1853–1916), Italian paleontologist, University of Naples.

Bassetite, locality at Basset group of mines, Redruth, Cornwall, England.

Bastnaesite, locality at Bastnäs, Vastmanland, Sweden.

Batisite, for composition, containing *Ba, Ti, Si*.

Baumhauerite, for Heinrich Adolf Baumhauer (1848–1926), German chemist and mineralogist, University of Freiburg.

Baumite, for John Leach Baum, American geologist, New Jersey Zinc Company.

Bauranoite, from composition, *Ba* and *uran*ium.

Bauxite (mixture of hydrated aluminum oxide minerals), locality at Baux, France.

Bavenite, locality at Baveno, Lago Maggiore, Piemonte, Italy.

Bayerite, apparently for Karl J. Bayer, nineteenth-century German metallurgist.

Bayldonite, for Dr. John Bayldon.

Bayleyite, for William Shirley Bayley (1861–1943), American geologist and mineralogist, University of Illinois and U.S. Geological Survey, editor of *Economic Geology.*

Baylissite, for Noel Stanley Bayliss, Austrailian chemist, University of Nedlands, Western Australia, who studied the synthetic compound in 1952.

Bazirite, from composition, contains *ba*rium and *zir*conium.

Bazzite, for Alessandro E. Bazzi, Italian engineer.

Bearsite, from composition, a *be*ryllium *ars*enate.

Beaverite, locality in Beaver County, Utah.

Becquerelite, for Antoine Henri Becquerel (1852–1908), French physicist who discovered radioactivity.

Behierite, for Jean Behier, mineralogist for the Service geologique, Madagascar.

Behoite, from composition, *Be* and *OH.*

Beidellite, locality at Beidell, Colorado.

Bellidoite, for Eleodoro Bellido Bravo, director of the Servicio de Geologia y Mineria, Peru.

Bellingerite, for H. C. Bellinger, Chile Exploration Company, who was associated with mining at the locality for the mineral.

Belovite, for Nikolai Vasilevich Belov (1891–   ), Russian crystallographer.

Belyankinite, for Dmitrii Stepanovich Belyankin (1876–1953), Russian mineralogist and petrographer.

Bementite, for Clarence Sweet Bement (1843–1923), American machine-tool manufacturer and collector of coins, books and minerals, Philadelphia.

Benitoite, locality in San Benito County, California.

Benjaminite, for Marcus Benjamin (1857–1932), American editor, U.S. National Museum, Washington, D.C.

Benstonite, for O. J. Benston (1901–   ), American ore dressing metallurgist, National Lead Company, Malvern, Arkansas, who furnished specimens for the initial study.

Beraunite, locality at Hrbek mine, near Beraun, in Bohemia.

Berborite, from composition, a *beryllium bor*ate.

Bergenite, locality near Bergen on the Trieb, Vogtland, Saxony.

Berlinite, for Nils Johan Berlin (1812–1891), Swedish pharmacologist, of the University of Lund.

Bermanite, for Harry Berman (1902–1944), American mineralogist, Harvard University.

Berndtite, for Fritz Berndt, of Corporation Minera de Bolivia, Oruro, Bolivia, who first recognized it as a new mineral.

Berryite, for Leonard Gascoigne Berry (1914–   ), Canadian mineralogist, Queen's University.

Berthierine, for Pierre Berthier (1782–1861), French geologist.

Berthierite, for Pierre Berthier (1782–1861), French geologist.

Bertossaite, for Antonio Bertossa, director of the Geological Survey of Rwanda, Africa.

Bertrandite, for Emile Bertrand, French mineralogist.

Beryl, from Greek *beryllos*, in ancient times applied to more than one kind of green mineral; its original significance is unknown.

Beryllite, from composition, a *beryll*ium mineral.

Beryllonite, from composition, a *beryll*ium mineral.

Berzelianite, for Jöns Jacob Berzelius (1779–1848), Swedish chemist.

Berzeliite, for Jöns Jacob Berzelius (1779–1848), Swedish chemist.

Beta-fergusonite, from Greek letter *beta* and fergusonite; dimorphous with fergusonite.

Betafite, locality at Betafo in Madagascar.

Beta-roselite, from Greek letter *beta* and roselite; dimorphous with roselite.

Beta-uranophane, from Greek letter *beta* and uranophane; dimorphous with uranophane.

Betekhtinite, for Anatolii Georgievich Betekhtin (1897–1962), Russian mineralogist and economic geologist.

Betpakdalite, locality in Bet-Pak-Dal desert, Central Kazakhstan, Soviet Union.

Beudantite, for François Sulpice Beudant (1787–1850), French mineralogist.

Beusite, for Alexei A. Beus, Russian mineralogist and geochemist, Moscow Polytechnic Institute.

Beyerite, for Adolph Beyer (1743–1805), German mining engineer and mineralogist, Schneeberg, Saxony.

Bianchite, for Angelo Bianchi (1892–   ), Italian mineralogist.

Bicchulite, locality at Bicchu, Okayama Perfecture, Japan.

Bideauxite, for Richard August Bideaux (1935–), American mineralogist, Tucson, Arizona, who first noted the mineral.

Bieberite, locality at Bieber in Hesse, Germany.

Bikitaite, locality in pegmatites of Bikita near Fort Victoria, Southern Rhodesia.

Bilinite, locality near Bilin, northern Bohemia.

Billietite, for Valère Billiet (1903–1944), Belgian crystallographer, University of Ghent.

Billingsleyite, for Paul Billingsley (1887–1962), American mining geologist, who collected the type specimens.

Bindheimite, for Johann Jacob Bindheim (1750–1825), German chemist, who made the first analysis of the mineral.

Biopyribole (collective name for these minerals), from the names *bio*tite (a mica), *pyr*oxene, and amph*ibole*.

Biotite, for Jean Baptiste Biot (1774–1862), French physicist, who studied the optical differences between the micas.

Biphosphammite, from composition, an *ammo*nium *biphosph*ate.

Biringuccite, for Vannoccio Biringuccio (1480–1539), Italian metallurgist and chemist.

Birnessite, locality at Birness, Scotland.

Bisbeeite, locality at Bisbee, Cochise County, Arizona.

Bischofite, for Gustav Bischof (1792–1870), German mineral chemist and geologist.

Bismite, from composition, contains *bism*uth.

Bismoclite, from composition, *bism*uth, *O,* and *Cl.*

Bismuth, apparently from German *Wismuth,* a term of uncertain origin; perhaps from old German for *meadow* and *claim to a mine,* or from German *Weisse Masse* (= *white mass*), or possibily from Greek for *lead white.*

Bismuthinite, from composition, a *bismuth* mineral.

Bismutite, from composition, a *bismut*h mineral.

Bismutoferrite, from composition, *bismut*h and iron (*ferr*um).

Bismutotantalite, from composition, *bismut*h and *tantal*um.

Bityite, locality on Mt. Bity, Madagascar.

Bixbyite, for Maynard Bixby, of Salt Lake City, Utah, who compiled a catalog of Utah minerals.

Bjarebyite, for Gunnar Bjareby (1899–1967), Swedish-American artist, naturalist, and mineral collector, Boston.

Blakeite, for William Phipps Blake (1826–1910), American geologist and mineralogist who worked in the American southwest.

Blende (obsolete term), from German *blenden,* meaning *to blind* or *deceive;* for example, applied to sphalerite because, while often resembling galena, it yielded no lead.

Blixite, for Ragner Blix (1898–   ), Swedish chemist, Swedish Museum of Natural History.

Bloedite, for Carl August Bloede (1773–1820), German chemist.

Bobierrite, for Pierre Adolphe Bobierre (1823–1881), French agricultural chemist.

Boehmite, for Johannes Böhm (1857–1938), German geologist and paleontologist, who first recognized the species through X-ray diffraction studies.

Bøggildite, for Ove Balthasar Bøggild (1872–?), Danish geologist, University of Copenhagen.

Bohdanowiczite, for Karol Bohdanowicz (1865–1947), Polish economic geologist, of Cracow.

Bokite, for Ivan Ivanovich Bok (1898–   ), Soviet geologist of Kazakhstan.

Boleite, locality at Boleo, Baja California, Mexico.

Bolivarite, for Ignacio Bolívar y Urrutía (1850–?), Spanish entomologist.

Boltwoodite, for Bertram Borden Boltwood (1870–1927), American radiochemist of Yale University.

Bonaccordite, locality in Bon Accord area, Barberton Mountain Land, Transvaal, Union of South Africa.

Bonattite, apparently for Stefano Bonatti (1902–1968), Italian petrologist, Pisa.

Bonchevite, for Georghi Bonchev, Bulgarian mineralogist, Institute of Mineralogy and Petrography, University of Sofia.

Boothite, for Edward Booth (1857–1917), American chemist, University of California.

Boracite, name derived from borax; and in allusion to its composition, a borate.

Borax, from the Persian *burah* and Arabic *buraq,* old names for the mineral.

Borcarite, from composition, a *bor*ate-*car*bonate.

Borickite, for Emanuel Borický (1840–1881), Czech petrographer.

Borishanskiite, for S. S. Borishanskaya, Soviet mineralogist.

Bornemanite, for Irina D. Borneman-Starynkevich, Russian mineralogist, student of minerals of the Khibina and Lovozero tundras.

Bornhardtite, for Wilhelm Bornhardt (1864–?), German student of ore deposits.

Bornite, for Ignaz von Born (1742–1791), Austrian mineralogist.

Borovskite, for Igor Borisovich Borovskii, Russian pioneer in electron microprobe analysis.

Botallackite, locality at Botallack mine, Cornwall, England.

Botryogen, from Greek words meaning *a bunch of grapeś* and *to bear*, in allusion to the botryoidal masses originally discovered.

Boulangerite, for Charles Louis Boulanger (1810–1849), French mining engineer.

Bournonite, for Jacques Louis de Bournon (1751–1825), French crystallographer and mineralogist.

Boussingaultite, for Jean Baptiste Boussingault (1802–1887), French chemist.

Bracewellite, for Smith Bracewell, director of the British Guiana Geological Survey, who first described the assemblage in which the mineral occurs.

Brackebuschite, for Luis Brackebusch (1849–1906), Argentine mineralogist, University of Córdoba.

Bradleyite, for Wilmot Hyde Bradley (1899–   ), American geologist, U.S. Geological Survey.

Braggite, for William Henry Bragg (1862–1942) and his son, William Lawrence Bragg (1890–1971), English physicists who pioneered in X-ray crystallography; the first new mineral isolated and determined by X-ray methods.

Braitschite, for Otto Braitsch (1921–1966), German geochemist and mineralogist, University of Freiburg.

Brammallite, for Alfred Brammall, British mineralogist, who studied clays.

Brandtite, for Georg Brandt (1694–1768), Swedish chemist.

Brannerite, for John Casper Branner (1850–1922), American geologist.

Brannockite, for Kent Combs Brannock (1923–1973), American chemist and mineralogist, Tennessee Eastman Company, Kingsport, Tennessee.

Brassite, for Rejane Brasse, who first synthesized the compound.

Braunite, for Kammerath Braun, of Gotha, Germany.

Bravoite, for José J. Bravo (1874–1928), Peruvian scientist of Lima.

Brazilianite, locality in Brazil, where the mineral was first found.

Bredigite, for Max Albrecht Bredig (1902–), physical chemist who studied polymorphism of calcium silicates.

Breithauptite, for Johann Friedrich August Breithaupt (1791–1873), German mineralogist, Saxony.

Brewsterite, for David Brewster (1781–1868), Scottish physicist who studied the optical properties of crystals.

Brezinaite, for Aristides Brezina (1848–1909), Austrian mineralogist, Natural History Museum, Vienna.

Brianite, for Brian Harold Mason (1917–), New Zealand-American geochemist and mineralogist, U.S. National Museum, Washington, D.C.

Briartite, for Gaston Briart, who studied type locality at Kipushi, Katanga.

Brindleyite, for George William Brindley (1905–), English-American mineralogist, Pennsylvania State University.

Britholite, from Greek for *weight,* in allusion to its relatively high specific gravity.

Brochantite, for André Jean François Marie Brochant de Villiers (1772–1840), French geologist and mineralogist.

Brockite, for Maurice R. Brock, American geologist, U.S. Geological Survey, who found the first specimens.

Bromargyrite, from composition, *brom*ium and silver (Greek, *argyr*os).

Bromellite, for Magnus von Bromell (1679–1731), Swedish physician and mineralogist.

Bronzite (var. of enstatite), named in allusion to its bronzelike appearance, especially on cleavage surfaces.

Brookite, for Henry James Brooke (1771–1857), English mineralogist.

Brownmillerite, for Lorrin Thomas Brownmiller (1902–), American chemist, Alpha Portland Cement Company, Easton, Pennsylvania.

Brucite, for Archibald Bruce (1777–1818), American mineralogist, who first described the species.

Brüggenite, for Juan Brüggen (1887–1953), Chilean geologist.

Brugnatellite, for Luigi Vincenzo Brugnatelli (1761–1818), Italian chemist, University of Pavia.

Brunogeierite, for Bruno H. Geier (1902–), mineralogist, Tsumeb Corporation, South West Africa.

Brushite, for George Jarvis Brush (1831–1912), American mineralogist, Yale University.

Buchwaldite, for Vagn Buchwald, Danish metallurgist, Danmarks Tekniske Højskole, Lyngby, who studied the meteorite in which the mineral was found.

Buddingtonite, for Arthur Francis Buddington (1890–), American geologist, Princeton University.

Buergerite, for Martin Julian Buerger (1903–), American crystallographer, Massachusetts Institute of Technology.

Buetschliite, for Otto Buetschli (1848–1920), German zoologist, University of Heidelberg, who studied the synthetic compound.

Bukovite, locality at Bukov, in western Bohemia.

Bukovskyite, for Antonin Bukovsky (1865–1950), Bohemian chemist, of the secondary school at Kutná Hora, Czechoslovakia.

Bultfonteinite, locality in the Bultfontein diamond mine, Kimberley, South Africa.

Bunsenite, for Robert Wilhelm Bunsen (1811–1899), German chemist, who also gave his name to a type of gas burner (in 1855).

Burangite, locality at the Burango pegmatite, Rwanda.

Burbankite, for Wilbur Sweet Burbank (1898–1975), American geologist, U.S. Geological Survey.

Burkeite, for William Edmund Burke (1880–   ), American chemical engineer, American Potash and Chemical Corporation, Trona, California, who had previously discovered the synthetic salt.

Bursaite, locality in the Bursa Province, Turkey.

Buserite, for Wilhelm Buser (1917–   ), University of Bern, Switzerland, who studied manganese minerals.

Bustamite, for General Anastasio Bustamente (1780–1853), of Mexico.

Butlerite, for Gurdon Montague Butler (1881–1961), American mining geologist, University of Arizona.

Buttgenbachite, for Henri Jean François Buttgenbach (1874–1964), Belgian mineralogist.

Byssolite (var. of actinolite or tremolite), from Greek for *flax* or *linen,* in allusion to the fibrous nature of the substance.

Bystomite, for Anders Byström, Swedish crystal chemist.

Bytownite, locality at Bytown (now the city of Ottawa), Canada.

C

Cacoxenite, from Greek words meaning *a bad guest,* because the phosphorus content injures the quality of the iron made from the limonite ore in which the mineral occurs.

Cadmoselite, from composition, contains *cadm*ium and *sel*enium.

Cadwaladerite, for Charles Meigs Biddle Cadwalader, American scientist, President of the Academy of Natural Sciences of Philadelphia.

Cafarsite, from composition, contains *ca*lcium, iron (*f*errum), *ars*enic, and others.

Cafetite, from composition, contains *Ca, Fe,* and *Ti.*

Cahnite, for Lazard Cahn (1865–1940), American mineral collector and dealer who first recognized the mineral at Franklin, New Jersey.

Cairngorm (var. of quartz), locality at Cairngorm, southwest of Banff, in Scotland.

Calamine (syn. for hemimorphite), perhaps from a corruption of Latin *cadmia*, ancient name for zinc silicate and carbonate, as well as zinc oxides from chimneys of furnaces; or maybe from Latin *calamus* (= *reed*), in allusion to the slender stalactitic forms common in furnace cadmia.

Calaverite, locality in the Stanislaus mine, Calaveras County, California.

Calciborite, from composition, *calci*um and *bor*on.

Calciocopiapite, from composition, like copiapite, but with predominant calcium.

Calcioferrite, from composition, *calci*um and *ferr*ic iron.

Calciouranoite, from composition, *calci*um and *urany*l.

Calciovolborthite, from composition, contains calcium and is chemically similar to volborthite.

Calcite, from Latin *calx* (= *lime*); the word *calcium* has the same origin, as does *chalk*.

Calcium-catapleiite, from composition, like catapleiite, but with predominant calcium.

Calcjarlite, from composition, like jarlite, but with predominant calcium.

Calclacite, from composition, *calc*ium, chlorine (*Cl*), and *ac*etate.

Calcurmolite, from composition, *calc*ium, *ur*anium, and *mol*ybdenum.

Caledonite, locality in Caledonia, the ancient name of Scotland, where it was first found.

Californite (var. of vesuvianite), locality in the state of California.

Calkinsite, for Frank Cathcart Calkins (1878–1974), American geologist, U.S. Geological Survey.

Callaghanite, for Eugene Callaghan (1904–   ), American geologist, University of Utah, Salt Lake City.

Calomel, perhaps from the Greek words for *beautiful* and *honey*, in allusion to the sweet taste (the *mercurius dulcis* of early chemistry); or maybe from Greek words for *beautiful* and *black*, in reference to black mercuric sulfide to which name was first applied.

Calumetite, locality at Calumet, Michigan.

Calzirtite, from composition, *calc*ium, *zir*conium, and *ti*tanium.

Canasite, from composition, *Ca*, *Na*, and *Si*.

Canavesite, locality at the village and mine of Brosso, Canavese district, Italy.

Cancrinite, for Count Georg Cancrin (1774–1845), German-Russian, Minister of Finance of Russia.

Canfieldite, for Frederick Alexander Canfield (1849–1926), American mining engineer and mineral collector of Dover, New Jersey.

Cannizzarite, for Stanislao Cannizzaro (1826–1910), Italian chemist, University of Rome.

Cappelenite, for D. Cappelen of Holden, Norway.

Caracolite, locality between Caracoles and the Bay of Mejillones, Chile.

Carboborite, from composition, contains *carbo*nate and *bor*ate groups.

Carbocernaite, from composition, *carbo*nate, *cer*ium, and sodium (*Na*).

Carbonate-apatite, from composition, the *carbonate* member of the apatite group.

Carbonate-cyanotrichite, from composition, like cyanotrichite, but with predominant carbonate.

Carbonate-fluorapatite, from composition, the carbonate-fluorine member of the apatite group.

Carbonate-hydroxylapatite, from composition, the carbonate-hydroxyl member of the apatite group.

Carbuncle (var. of members of the garnet group), from Latin *carbunculus* (= *little ember*), in allusion to the internal firelike color and reflection.

Carletonite, for Carleton University, Ottawa, Canada, where it was first recognized.

Carlfriesite, for Carl Fries, Jr. (1910–1965), geologist of the U.S. Geological Survey and the Instituto de Geologia of the National University, Mexico.

Carlinite, locality at the Carlin gold deposit, northern Eureka County, Nevada.

Carlsbergite, for the Carlsberg Foundation, Copenhagen, which supported research on the mineral.

Carminite, from carmine, in allusion to the carmine-red color.

Carnallite, for Rudolph von Carnall (1804–1874), Prussian mining engineer.

Carnelian (var. of quartz), perhaps from the Latin *carneus,* meaning *fleshy,* in allusion to the color; or perhaps from medieval Latin name *cornus* for a species of dogwood with a reddish berry.

Carnotite, for Marie Adolphe Carnot (1839–1920), French mining engineer and chemist.

Carobbiite, for Guido Carobbi (1900–), Italian geologist, University of Florence.

Carpholite, from Greek word for *straw,* in allusion to its yellow color.

Carrboydite, locality at the Carr Boyd mine, Western Austrailia.

Carrollite, locality in Carroll County, Maryland.

Caryinite, from Greek for *nut-brown,* in allusion to its color.

Caryopilite, from Greek for *nut* and *felt,* in allusion to its brown color and felted structure.

Cassidyite, for William A. Cassidy, who mapped the Wolf Creek crater in Western Australia, where mineral was found.

Cassiterite, from Greek *kassiteros* (= *tin*).

Catapleiite, from Greek for *wholly* and *full,* because it is always accompanied by a number of rare minerals.

Cattierite, for Felicien Cattier (1869–1946), Chairman of the Board, Union Minière du Haut-Katanga, Africa.

Cavansite, from composition, *ca*lcium, *van*adium, and *si*licate.

Caysichite, from composition, contains *Ca, Y, Si, C, H.*

Cebollite, locality at Cebolla Creek, Gunnison County, Colorado.

Celadonite, from French *celadon* (= *sea green*), in allusion to the color of the mineral.

Celestite, from Latin *coelestis* (= *heavenly*), in allusion to the faint blue color of the first specimens described.

Celsian, for Anders Celsius (1701–1744), Swedish astronomer and naturalist.

Cerargyrite (syn. for chlorargyrite), from Greek for *horn* and for *silver,* in allusion to its hornlike appearance and composition.

Cerianite, from composition, *ceri*um dioxide.

Cerite, for the asteroid Ceres, discovered in 1801, two years before the mineral and the element cerium were named; the element was discovered in this mineral.

Cernyite, for Petr Cerný (1934–   ), Czech-Canadian geologist, University of Manitoba, Winnipeg, who provided some of the original specimens.

Cerotungstite, from composition, *ce*rium and *tungst*en, and similarity to yttrotungstite.

Ceruleite, from *cerulean* blue, in allusion to its color.

Cerussite, from an ancient Greek term applied to lead carbonate, which became *cerussa* in Latin.

Cervantite, locality at Cervantes, Galicia, Spain.

Cesarolite, for Giuseppe R. P. Césaro (1849–1939), Belgian mineralogist, University of Liège.

Cesbronite, for Fabien Cesbron, French mineralogist.

Cesium kupletskite, from composition, like kupletskite, but with predominant cesium.

Chabazite, from Greek *chabazios* or *chalazios,* an ancient name of a stone; the name of the last of the twenty stones celebrated for their virtues and mentioned in a poem ascribed to Orpheus.

Chalcanthite, from Greek for *copper* and *flower,* in allusion to its common efflorescent forms.

Chalcedony (var. of quartz), from Chalcedon or Calchedon, an ancient maritime city of Bithynia, on the Sea of Marmara, Asia Minor.

Chalcoalumite, from composition, copper (*chalkos*) and *alum*inum.

Chalcocite, from composition, copper (*chalkos*).

Chalcocyanite, from Greek for *copper* and *azure-blue,* in allustion to composition and color.

Chalcomenite, from Greek for *copper* and *moon,* in allusion to its content of copper and selenium (element name is derived from a different Greek word meaning moon).

Chalconatronite, from composition, copper (*chalkos*) and sodium (*natrium*).

Chalcophanite, from Greek for copper and *to appear,* in allusion to the change of color on ignition.

Chalcophyllite, from Greek for copper and word for *a leaf,* becaue of its micaceous structure and copper content.

Chalcopyrite, from composition, a copper (*chalkos*) bearing mineral similar to pyrite in appearance.

Chalcosiderite, from composition, copper (*chalkos*) and iron (*sideros*).

Chalcostibite, from composition, copper (*chalkos*) and antimony (*stib*ium).

Chalcothallite, from composition, copper (*chalkos*) and *thall*ium.

Chalcotrichite (var. of cuprite), from Greek for copper and for *hair,* in allusion to its copper content and capillary structure.

Challantite, locality at the gold mine of Challant-St. Anselme (Valle d'Ayas), Aosta, Italy.

Chambersite, locality in Chambers County, Texas.

Chamosite, locality at Chamoson, near St. Maurice, in Valais Canton, Switzerland.

Chaoite, for Edward Ching-Te Chao (1919–   ), Chinese-American astrogeologist, U.S. Geological Survey.

Chapmanite, for Edward John Chapman (1821–1904), Canadian geologist and mineralogist, University of Toronto.

Charoite, locality at the Charo River area in the Murun massif, northwest Aldan, Yakutsk, Soviet Union.

Chelkarite, locality near Chelkar, Kazakhstan, Soviet Union.

Chenevixite, for Richard Chenevix (1774–1830), Irish-French chemist and mineralogist.

Cheralite, from Chera, the ancient Dravidian kingdom, which corresponds roughly to the modern territory of Travancore in India.

Chernovite, for Aleksandr Aleksandrovich Chernov (1877–?), Russian geologist.

Chernykhite, for V. V. Chernykh, Russian professor, Leningrad Mining Institute.

Chervetite, for Jean Chervet, French mineralogist.

Chesterite, locality near Chester, Vermont.

Chevkinite, for General Konstantin Vladimirovich Tschevkin [Chevkin] (1802–1875), Chief of the Russian Mining Corps.

Chiastolite (var. of andalusite), from Greek for *arranged diagonally*, and hence from *chi*, the Greek letter resembling X, in allusion to the cross exhibited in transverse sections.

Childrenite, for John George Children (1777–1852), English chemist and mineralogist.

Chiolite, from Greek for *snow*, in allusion to its appearance and similarity to cryolite (from Greek for *ice*).

Chkalovite, for Valery Pavlovich Chkalov (1904–1938), Russian aviator, first to make a nonstop flight from Moscow, via the North Pole, to the United States.

Chloanthite, from Greek for *budding*, in allusion to the green color of annabergite, its alteration product.

Chloraluminite, from composition, *chlor*ine and *alumin*um.

Chlorapatite, from composition, the *chlor*ine member of the apatite group.

Chlorargyrite, from composition, *chlor*ine and silver (*argyr*os).

Chlorite (mineral group name), from a Greek word meaning *green*, in allusion to the common color.

Chloritoid, name derived from chlorite, because of its superficial resemblance to the minerals of that group.

Chlormanganokalite, from composition, *chlor*ide, *mangano*us, potassium (*kal*ium).

Chlorophoenicite, from Greek for *green* and *purple-red*, in allusion to its different colors under natural and artificial light.

Chlorothionite, from composition, *chlor*ine and sulfur (Greek *theio*n).

Chloroxiphite, from Greek for *green* and *blade* or *straight sword*, in allusion to color and the habit of the crystals.

Chondrodite, from Greek for *granule*, alluding to its occurrence as isolated grains.

Christite, for Charles Louis Christ (1916–   ), American physical chemist, U.S. Geological Survey.

Chromatite, from composition, a *chroma*te.

Chromite, from composition, contains *chrom*ium.

Chrysoberyl, from Greek for *golden* and the mineral beryl, in allusion to its color and beryllium content.

Chrysocolla, from Greek words meaning *gold* and *glue,* originally given by the ancients to a mineral or minerals used for soldering gold, but long applied to various green copper minerals.

Chrysoprase (var. of quartz), from Greek for *gold* and for *leek,* in allusion to its similarity to prase (from Greek meaning *leek green*).

Chrysotile, from Greek for *gold* and *fiber,* in allusion to its color and asbestiform nature.

Chudobaite, for Karl Franz Chudoba (1898–1968), German mineralogist.

Chukhrovite, for Fedor Vasilevich Chukhrov (1908–), Russian mineralogist.

Churchite, for Arthur Herbert Church (1834–1915), English chemist.

Cinnabar, from Medieval Latin *cinnabaris;* with a long history which can be traced to Persian *zinjifrah,* apparently meaning *dragon's blood,* in allusion to its red color.

Citrine (var. of quartz), from French *citron* (= *lemon*), in allusion to its yellow color.

Claringbullite, for Gordon Frank Claringbull (1911–), British mineralogist, director of the British Museum (Natural History), London.

Clarkeite, for Frank Wigglesworth Clarke (1847–1931), American geochemist of the U.S. Geological Survey.

Claudetite, for Frederick Claudet, French chemist who first described the mineral.

Clausthalite, locality at Clausthal, Germany.

Cleavelandite (var. of albite), for Parker Cleaveland (1780–1858), American mineralogist, Bowdoin College, Brunswick, Maine.

Cliffordite, for Clifford Frondel (1907–), American mineralogist, Harvard University.

Cliftonite, for Robert Bellamy Clifton (1836–1921), British physicist, Oxford.

Clinobisvanite, from crystal system, mono*clin*ic, and composition, contains *bis*muth and *van*adium.

Clinochlore, from crystal system, mono*clin*ic, and Greek for *green,* in allusion to its color.

Clinochrysotile, from crystal system, the mono*clin*ic dimorph of chrysotile.

Clinoclase, from Greek for *to slant* and *to break,* in allusion to the oblique basal cleavage.

Clinoenstatite, from crystal system, the mono*clin*ic dimorph of enstatite.

Clinoferrosilite, from crystal system, the mono*clin*ic dimorph of ferrosilite.

Clinohedrite, from Greek words for *to slant* and *face,* from the prevalence of forms without parallel faces in the monoclinic domatic class in which it crystallizes.

Clinoholmquistite, from crystal system, the monoclinic dimorph of holmquistite.

Clinohumite, from crystal system, monoclinic, and its position in the humite group.

Clinohypersthene, from crystal system, the monoclinic dimorph of hypersthene.

Clinojimthompsonite, from crystal system, the monoclinic dimorph of jimthompsonite.

Clinoptilolite, from crystal system, monoclinic, and Greek for *wing* or *down,* in allusion to the light, downy nature of aggregates of the mineral.

Clinosafflorite, from crystal system, the monoclinic dimorph of safflorite.

Clinoungemachite, from crystal system, monoclinic, and chemical similarity to ungemachite.

Clinozoisite, from crystal system, the monoclinic dimorph of zoisite.

Clintonite, for DeWitt Clinton (1769–1828), American lawyer and statesman who was interested in geology and other sciences.

Coalingite, locality near Coalinga, California.

Cobaltite, from composition, it contains *cobalt.*

Cobaltomenite, from composition, *cobalt,* and Greek for *moon,* in allusion to its content of selenium (element name is derived from a different Greek word meaning moon).

Cobalt pentlandite, from composition, like pentlandite, but with predominant cobalt.

Cobalt-zippeite, from composition, like zippeite, but with predominant cobalt.

Coconinoite, locality in Coconino County, Arizona.

Coeruleolactite, from Latin *coeruleus* (= *blue*), and *lactis* (= *milk*), in allusion to its milk white to light blue color.

Coesite, for Loring Coes, Jr. (1915–), American research chemist, Norton Company, Worcester, Massachusetts, who first synthesized the compound.

Coffinite, for Reuben Clare Coffin (1886–1972), American geologist, Tulsa, Oklahoma, a pioneer in the study of uranium deposits in the Colorado Plateau.

Cohenite, for Emil Wilhelm Cohen (1842–1905), German mineralogist, Greifswald.

Colemanite, for William Tell Coleman (1824–1893), American merchant, San Francisco, California, founder of the California borax industry.

Collinsite, for William Henry Collins (1878–1937), Canadian geologist, director of the Geological Survey of Canada.

Collophane (general term for fine-grained members of the apatite group), from Greek for *glue* and *to appear,* in allusion to its colloidal aspect.

Coloradoite, locality in the state of Colorado.

Columbite, from Columbia, an old name for the United States of America, where the original specimen was obtained; in that specimen the element columbium (= niobium) was first recognized in 1802.

Colusite, locality near the Colusa claim, Butte, Montana.

Combeite, for Arthur Delmar Combe of the Geological Survey of Uganda.

Compreignacite, locality near Compreignac, France.

Congolite, locality near Brazzaville, Congo.

Conichalcite, from Greek for *lime* and *copper,* in allusion to the presence of both calcium and copper.

Connellite, for Arthur Connell (1794–1863), Scottish chemist, University of St. Andrews, who first examined the mineral.

Cookeite, for Josiah P. Cooke, Jr. (1827–1894), American chemist and mineralogist, Harvard University.

Cooperite, for R. A. Cooper, who first described the mineral.

Copiapite, locality near Copiapó, Chile.

Copper, derived from Latin *cuprum,* which in turn came from Greek *kyprios* (for Cyprus, where the metal was early found).

Coquimbite, locality in the province of Coquimbo, Chile.

Corderoite, locality at Cordero mine, Humboldt County, Nevada.

Cordierite, for Pierre Louis A. Cordier (1777–1861), French mining engineer and geologist.

Cordylite, from Greek word for *club,* in allusion to the shape of the scepterlike crystals.

Corkite, locality in the county of Cork, Ireland.

Cornetite, for Jules Cornet (1865–1929), Belgian geologist.

Cornubite, from Cornubia, the old Roman name for Cornwall in England, the locality of the mineral.

Cornwallite, locality at Cornwall, England.

Coronadite, for Francisco Vasquez de Coronado (ca. 1500–1554), Spanish explorer of the American southwest.

Corrensite, for Carl Wilhelm Correns (1893–   ), German mineralogist, director of the Sedimentary Petrography Institute, Göttingen University.

Corundophilite, from mineral corundum and the Greek word for *friend,* because of its association with corundum in the type locality.

Corundum, apparently derived from the old Indian term *kauruntaka,* a name for the mineral.

Corvusite, from Latin *corvus* (= *raven*), in allusion to its blue-black color.

Cosalite, locality at the Cosala mine, Sinaloa, Mexico.

Costibite, from composition, *co*balt and antimony (*stib*ium).

Cotunnite, for Domenico Cotugno [Cotunnius] (1736–1822), Italian physician and anatomist, University of Naples.

Coulsonite, for Arthur Lennox Coulson (1898–   ), of the Indian Geological Survey.

Cousinite, for Jules Cousin (1884–1965), president of the board of directors and member of the administrative council of l'Union Minière du Haut-Katanga.

Covellite, for Niccolò Covelli (1790–1829), Italian mineralogist, who discovered Vesuvian covellite.

Cowlesite, for John Cowles, American amateur mineralogist and zeolite collector, of Rainier, Oregon.

Crandallite, for Milan L. Crandall, Jr., American engineer, Knight Syndicate, Provo, Utah.

Creaseyite, for Saville Cyrus Creasey (1917–   ), American geologist, U.S. Geological Survey, who studied Arizona mines.

Crednerite, for Karl F. H. Credner (1809–1876), German mining geologist and mineralogist.

Creedite, locality in the Creede quadrangle, Mineral County, Colorado.

Crichtonite, for Alexander Crichton (1763–1856), Scottish physician and mineral collector; with Alexander I in Russia for a time.

Cristobalite, locality of Cerro San Cristóbal near Pachuca, Mexico.

Crocidolite (var. of riebeckite), from Greek for *woof,* in allusion to its asbestiform structure.

Crocoite, from Greek for *saffron,* in allusion to saffron orange color of the mineral.

Cronstedtite, for Axel Fredrik Cronstedt (1722–1765), Swedish mineralogist and chemist.

Crookesite, for William Crookes (1832–1919), English chemist and physicist, who in 1861 discovered the element thallium, which is contained in the mineral.

Crossite, for Charles Whitman Cross (1854–1949), American geologist, U.S. Geological Survey.

Cryolite, from Greek for *ice* or *frost,* in allusion to the icy appearance of the mineral.

Cryolithionite, from mineral cryolite and its lithium content, in allusion to its similarity to the composition of cryolite.

Cryptohalite, from Greek for *concealed* and *salt,* in allusion to its intimate mixture with other salts in the original locality.

Cryptomelane, from Greek words for *concealed* and *black,* because the identity of the mineral as a distinct species was hidden in the ill-defined psilomelane minerals.

Csiklovaite, locality at Csiklova [Ciclova], Rumania.

Cubanite, locality in Cuba, the country in which the mineral was first found.

Cumengite, for Edouard Cumenge (1828–1902), French mining engineer.

Cummingtonite, locality at Cummington, Massachusetts.

Cuprite, from Latin *cuprum* (= *copper*), in allusion to its composition.

Cuprobismutite, from composition, copper (*cupr*um) and *bismut*h.

Cuprocopiapite, from composition, like copiapite, but with predominant copper (*cupr*um).

Cuprorivaite, from composition, copper (*cupr*um), and for Carol Riva (1872–1902), mineralogist.

Cuprosklodowskite, from composition, like sklodowskite, but with predominant copper (*cupr*um).

Cuprospinel, from composition, copper (*cupr*um), and the fact it is in the spinel group.

Cuprostibite, from composition, contains copper (*cupr*um), and antimony (*stib*ium).

Cuprotungstite, from composition, contains copper (*cupr*um) and *tungst*en.

Curienite, for Hubert Curien (1924–   ), French crystallographer and mineralogist, Centre national de la recherche scientifique, Paris.

Curite, for Pierre Curie (1859–1906), French physicist known for his research on radioactivity.

Cuspidine, from Latin *cuspis* (= *spear*), in allusion to the characteristic form of the crystals.

Cyanochroite, from Greek for *blue* and *color,* in allusion to the color of the mineral.

Cyanotrichite, from Greek for *blue* and *hair,* in allusion to the color and the fibrous nature of the mineral.

Cyclowollastonite, from the Greek word for *circle,* and the mineral wollastonite, in allusion to the fact that this mineral has a cyclosilicate structure while wollastonite has an inosilicate structure.

Cylindrite, from Greek word for *roll,* in allusion to the cylindrical form of the mineral.

Cymrite, from Cymru, the old Welsh name for Wales, the locality of the mineral.

Cyrilovite, locality in pegmatite at Cyrilov, near Velké Mežirici, West Moravia, Czechoslovakia.

Cyrtolite (var. of zircon), from Greek word for *bent,* in allusion to the strongly curved pyramidal faces on some of the original specimens.

D

Dachiardite, for Antonio d'Achiardi (1839–1902), Italian mineralogist, the mineral having been investigated by his son.

Dadsonite, for Alexander Stewart Dadson (1906–1958), who contributed to the development of mining in the Yellowknife gold deposits, Northwest Territories, Canada.

Dahllite (syn. for carbonate-hydroxylapatite), for Tellef Dahll (1825–1893) and Johan Martin Dahll (1830–1877), Norwegian geologists and brothers.

Dalyite, for Reginald Aldworth Daly (1871–1957), American geologist, Harvard University.

Danaite (syn. for cobaltoan arsenopyrite), for James Freeman Dana (1793–1827) of Boston, who first made known the Franconia, New Hampshire occurrence.

Danalite, for James Dwight Dana (1813–1895), American mineralogist and geologist, Yale University.

Danburite, locality at Danbury, Connecticut.

D'Ansite, for Jean D'Ans (1881–), German chemist, professor, Berlin.

Daomanite (= taomaite), locality in Tao and Ma districts of China.

Darapiosite, locality at Dara-Pioz, northern Tadzhikistan, Soviet Union.

Darapskite, for Ludwig Darapsky (1857–?), Chilean mineralogist and chemist, of Santiago.

Dashkesanite, locality at Dashkesan, Transcaucasia.

Datolite, from Greek for *to divide,* on account of the granular character of some of its varieties.

Daubreeite, for Gabriel Auguste Daubrée (1814–1896), French geologist of Paris.

Daubreelite, for Gabriel Auguste Daubrée (1814–1896), French geologist of Paris.

Davidite, for Tennatt William Edgeworth David (1858–1934), Australian geologist.

Davisonite, for John Mason Davison (1840–1915), American chemist and mineralogist, University of Rochester, who studied meteorites.

Davreuxite, for Charles Joseph Davreux (1800–1863), Belgian pharmacist and natural scientist who contributed to mineralogy, geology, and chemistry.

Davyne, for Sir Humphry Davy (1778–1829), English chemist.

Dawsonite, for John William Dawson (1820–1899), Canadian geologist and principal of McGill University.

Dayingite (= tayingite), for Chinese type locality, no specific name given.

Deerite, for William Alexander Deer (1910–   ), British mineralogist and petrologist, Cambridge University.

Dehrnite, locality at Dehrn, Nassau, Germany.

Delafossite, for Gabriel Delafosse (1796–1878), French mineralogist and crystallographer.

Delhayelite, for Fernard Delhaye (1880–1946), Belgian gelogist.

Dellaite, for Della Martin Roy (1926–   ), Pennsylvania State University, University Park.

Delrioite, for Andrés Manuel del Rio (1764–1849), Mexican mineralogist, who first found vanadium in North America.

Delvauxite, for J. S. P. J. Delvaux de Feuffe (1782–?), Belgian chemist, who first analyzed the mineral.

Demantoid (var. of andradite), from German *Demant* (= *diamond*), in reference to its high luster.

Demesmaekerite, for M. G. Demesmaeker, director of the geological department of Union Minière du Haut Katanga in Africa.

Denningite, for Reynolds McConnell Denning (1916–1967), American mineralogist, University of Michigan, Ann Arbor.

Derbylite, for Orville Adelbert Derby (1851–1915), American geologist, Director of the Geological Survey of Brazil.

Derriksite, for Joseph Derriks, geologist, who studied the Shinkolobwe uranium deposits of Africa.

Descloizite, for Alfred Louis Olivier Legrand Des Cloizeaux (1817–1897), French mineralogist, who first described the mineral.

Despujolsite, for Pierre Despujols (1888–   ), of the Service Carte Géologique, Morocco.

Devilline, for Henri Etienne Sainte Claire Deville (1818–1881), French chemist.

Dewindtite, for Jean Dewindt, Belgian geologist.

Diabantite, modified form of the Greek term, *diabantachronnyn,* meaning *to color* or *stain diabase;* the mineral imparting a green color to diabase rock.

Diaboleite, from Greek for *apart* or *distinct from* and the mineral boleite.

Diadochite, from Greek for *a successor,* probably for the fact it forms by the alteration of previously formed phosphates.

Diallage, from Greek word for *difference,* in allusion to the dissimilar way it breaks, due to parting, from other, similar pyroxenes.

Diamond, from corruption of the Greek word *adamas,* meaning *invincible,* alluding to its hardness and supposed indestructibility.

Diaphorite, from Greek word for *difference;* although similar it is different from freieslebenite.

Diaspore, from Greek for *to scatter,* in allusion to its decrepitation when heated.

Dickinsonite, for Rev. John Dickinson of Redding, Connecticut, who collected rare minerals from the locality of the mineral.

Dickite, for Allan Brugh Dick (1833–1926), Scottish metallurgical chemist.

Dienerite, for Karl Diener (1862–1928), Austrian paleontologist, of Vienna, who discovered the mineral.

Dietrichite, for G. W. Dietrich, chemist of Pribram, Bohemia, who analyzed the mineral.

Dietzeite, for August Dietze (?–1893?), who first described the mineral.

Digenite, from Greek for *two sexes* or *kinds,* because it supposedly contained both cupric and cuprous ions.

Dimorphite, from Greek word meaning *two forms;* a supposed dimorph with the mineral realgar.

Diopside, from Greek for *two* and *view,* since the vertical prism zone apparently can be oriented in two ways.

Dioptase, from Greek for *through* and *to see,* because cleavage directions were distinguishable on looking through the crystal.

Dipyre (var. of scapolite), from Greek for *double* and *fire,* for the two heating properties of fusion and phosphorescence.

Dittmarite, apparently for William Dittmar (1833–1892), Scottish chemist of Glasgow.

Dixenite, from Greek for *two* and *stranger,* from the unique association of silica and arsenious oxide in the same compound.

Djerfisherite, for Daniel Jerome Fisher (1896–   ), American mineralogist, University of Chicago.

Djurleite, for S. Djurle, chemist, who made the synthetic compound before it was found in nature.

Dolerophanite, from Greek words for *fallacious* and *to appear,* because its brown color does not suggest a copper mineral.

Dolomite, for Deodat Guy Silvain Tancrède Gratet de Dolomieu (1750–1801), French geologist and mineralogist.

Doloresite, locality near Dolores River, a major stream of the Colorado Plateau, in Colorado.

Domeykite, for Ignacio Domeyko (1802–1889), Chilean mineralogist.

Donathite, for Martin Donath (1901–1965), German economic geologist, Technical University, Berlin.

Donnayite, for Joseph Désiré Hubert Donnay (1902–), Belgian-American mineralogist and crystallographer, Johns Hopkins and McGill Universities, and his wife Gabrielle (Hamburger) Donnay (1920–), crystallographer and mineralogist, McGill University, Montreal.

Douglasite, locality at Douglashall, northwest of Stassfurt, Saxony.

Downeyite, for Wayne F. Downey, Jr., of Harrisburg, Pennsylvania, who collected the original specimens.

Dravite, locality in district of Drave, Carinthia, Austria.

Dresserite, for John Alexander Dresser (1866–1954), Canadian geologist, contributed to the geology of the Monteregian Hills.

Drysdallite, for A. R. Drysdall, Director, Geological Survey Department of Zambia.

Dufrenite, for Ours Pierre Armand Petit Dufrenoy (1792–1857), French mineralogist and geologist, Ecole des Mines, Paris.

Dufrenoysite, for Ours Pierre Armand Petit Dufrenoy (1792–1857), French mineralogist and geologist, Ecole des Mines, Paris.

Duftite, for G. Duft, a director of the mines at Tsumeb, South West Africa.

Dumontite, for André Hubert Dumont (1809–1857), Belgian geologist.

Dumortierite, for Eugène Dumortier (1802–1873?), French paleontologist, of Lyons.

Dundasite, locality at Dundas, Tasmania.

Durangite, locality in state of Durango, Mexico.

Duranusite, locality at Duranus, Alpes-Maritimes, France.

Dussertite, for Désiré Dussert (1872–?), French mining engineer, studied Algerian deposits.

Duttonite, for Clarence Edward Dutton (1841–1912), American geologist, U.S. Geological Survey, one of first geologists to work in Colorado Plateau region.

Dypingite, locality at Dypingdal, Snarum, southern Norway.

Dyscrasite, from Greek for *bad mixture,* alluding to the combination of antimony with silver.

Dzhalindite, locality in Dzhalind ore deposit, Little Khingan Ridge, Far Eastern Siberia, Soviet Union.

Dzhezkazganite, locality in the Dzhezkazgan deposits, Kazakhstan, Soviet Union.

E

Eakerite, for Jack Eaker of Kings Mountain, North Carolina, who first noticed the mineral.

Eardleyite, for Armand John Eardley (1901–1972), American geologist, University of Utah.

Earlandite, for Arthur Earland, English civil servant.

Ecdemite, from Greek for *unusual*, in allusion to the composition, an oxychloride of lead and arsenic.

Eckermannite, for Claes Walther Harry von Eckermann (1886–1969), Swedish petrologist of Stockholm.

Edenite, locality at Edenville, New York.

Edingtonite, for Mr. Edington of Glasgow, Scotland, who discovered the mineral.

Eglestonite, for Thomas Egleston (1832–1900), American mineralogist and metallurgist, Columbia University, New York.

Eitelite, for Wilhelm Eitel (1891–   ), German-American chemist, Institute for Silicate Research, University of Toledo, Ohio.

Ekanite, for F. L. D. Ekanayake, who first found the mineral in Ceylon.

Elbaite, locality on the island of Elba off the coast of Italy.

Electrum, from Greek and Latin for *amber*, in allusion to the yellow color of the substance.

Eleolite (var. of nepheline), from Greek for *oil*, in allusion to its greasy luster.

Ellestadite, for Reuben B. Ellestad (1900–   ), American analytical chemist of Minneapolis, Minnesota.

Ellsworthite (var. of betafite), for Hardy Vincent Ellsworth (1889–1952), Canadian mineralogist, Canadian Geological Survey.

Elpasolite, locality in El Paso County, Colorado, where it was discovered.

Elpidite, from Greek for *hope*, because there was expectancy of finding other interesting minerals from the same locality.

Elyite, for John Ely, nineteenth-century mining promoter, an important figure in the mining history of eastern Nevada.

Embolite, from Greek for *intermediate*, because intermediate between chloride and bromide of silver.

Embreyite, for Peter Godwin Embrey (1929–   ), English mineralogist, British Museum of Natural History, London.

Emeleusite, for Charles Henry Emeleus (1930–   ), English petrologist, University of Durham, England.

Emerald (var. of beryl), an ancient term applied to various green minerals; from Latin *smaragdus* and Greek *smaragdos,* probably of Semitic origin.

Emery (var. of corundum), a term with a long history, from French *emeri,* Italian *smeriglio,* and Greek *smiris* or *smeris;* akin to Greek *myron* (= *unguent*).

Emmonsite, for Samuel Franklin Emmons (1841–1911), American geologist, U.S. Geological Survey.

Emplectite, from Greek for *entwined* or *interwoven,* in allusion to its intimate association with quartz.

Empressite, locality at the Empress Josephine mine, Kerber Creek district, Saguache County, Colorado.

Enargite, from Greek for *distinct,* referring to its perfect cleavage.

Endellite, for Kurd Endell (1887–   ), German geologist, one of its discoverers.

Endlichite (var. of vanadinite), for Frederick Miller Endlich (1851–1899), American geologist, superintendent of the mine in New Mexico where the mineral was discovered.

Englishite, for George Letchworth English (1864–1944), American mineral dealer and collector.

Enstatite, from Greek for *opponent,* because of its refractory nature before the blowpipe.

Eosphorite, from Greek for *dawn-bearing,* in allusion to the pink color of the first specimens studied.

Ephesite, locality at Gumuch-dagh, near Ephesus, Asia Minor.

Epididymite, from Greek for *near* and *-didymite,* in allusion to the fact that it is dimorphous with eudidymite.

Epidote, from Greek for *increase,* because the base of the prism has one side longer than the other.

Epigenite, from Greek for *to follow after,* because it is always observed implanted upon other minerals.

Epistilbite, from Greek for *near* and stilbite, a mineral which is similar in many respects.

Epistolite, from Greek word for *letter,* in allusion to the flat rectangular form and white color.

Epsomite, locality at Epsom, England.

Ericaite, for its violet color, which resembles the flower of heather (genus *Erica*).

Ericssonite, for John E. Ericsson (1803–1889), Swedish-American inventor and engineer, designer of the iron-clad ship *Monitor.*

Eriochalcite, from Greek for *wool* and *copper,* because it is a fibrous copper mineral.

Erionite, from Greek for *wool,* in allusion to its white, fibrous, woollike appearance.

Erlichmanite, for Jozef Erlichman, electron microprobe analyst, Palo Alto, California.

Ernstite, for Theodor K. H. Ernst (1904–   ), German mineralogist, Erlangen University.

Erythrite, from Greek for *red,* in allusion to color of the mineral.

Erythrosiderite, from Greek for *red* and *iron,* in allusion to its color and composition.

Eskebornite, locality at Eskeborn adit, Harz Mountains, Germany.

Eskimoite, for the Eskimo people, early settlers of Greenland, the original locality for the mineral.

Eskolaite, for Pentti Eelis Eskola (1883–1964), Finnish geologist, University of Helsinki.

Esperite, for Esper Signius Larsen, Jr. (1879–1961), American petrologist, Harvard University.

Ettringite, locality near Ettringen, Rhine Province, Germany.

Eucairite, from Greek for *opportunely,* because it was found soon after the discovery of selenium, which it contains.

Euchlorin, from Greek for *pale green,* in allusion to its color.

Euchroite, from Greek for *beautiful color,* alluding to its bright, emerald green color.

Euclase, from Greek for *good* and *fracture,* in allusion to the easy cleavage.

Eucryptite, from Greek for *well* and *concealed,* alluding to its mode of occurrence, often embedded in albite.

Eudialyte, from Greek for *good* and *to dissolve,* alluding to its easy solubility in acids.

Eudidymite, from Greek for *good* and *twin,* in allusion to its occurrence in twinned crystals.

Eulytite, from Greek for *good* and *dissolved,* in allusion to its easy fusibility.

Euxenite, from Greek for *friendly to strangers* or *hospitable,* in allusion to the rare elements it contains.

Evansite, for Brooke Evans (1797–1862), English nickel refiner of Birmingham, who brought the first specimens from Hungary.

Eveite, from the name Eve, in allusion to the close relationship of the mineral to adamite, with which it is isostructural.

Evenkite, locality in Evenki region, Lower Tunguska River, Siberia, Soviet Union.

Ewaldite, for Paul Peter Ewald (1888–   ), German-American physicist and crystallographer, Polytechnic Institute of New York.

Eylettersite, named for the wife of the discoverer, L. Van Wambeke.

Ezcurrite, for Juan Manuel de Ezcurra, manager of the Compañía Productora de Boratos, Salta province, Argentina.

F

Fabianite, for Hans Joachim Fabian, German geologist.

Faheyite, for Joseph John Fahey (1901–   ), American geochemist, U.S. Geological Survey, Washington, D.C.

Fairchildite, for John Gifford Fairchild (1882–1965), American geochemist, U.S. Geological Survey, Washington, D.C.

Fairfieldite, locality in Fairfield County, Connecticut.

Falcondoite, for the mining company, Falconbridge Dominicana C. por A. (Falcondo), which mines laterite at the type locality near Bonao, Dominican Republic.

Famatinite, locality at Sierra de Famatina, Argentina.

Farringtonite, for Oliver Cummings Farrington (1864–1933), Curator of Geology, Field Museum, Chicago, Illinois.

Fassaite, locality in the Fassathal, Tyrol.

Faujasite, for Barthélemy Faujas de Saint Fond (1741–1819), French geologist, writer on origin of volcanoes.

Faustite, for George Tobias Faust (1908–   ), American mineralogist and geologist, U.S. Geological Survey, Washington, D.C.

Fayalite, locality on Fayal Island in the Azores where it was believed to have occurred in a local volcanic rock; it was probably obtained from slag carried as ship's ballast.

Fedorite, for Evgraf Stepanovich Fedorov (1853–1919), Russian mineralogist and crystallographer.

Fedorovskite, for Nikolai Mikhailovich Fedorov (1886–1956), Russian mineralogist, first director of the All-Union Scientific Research Institute of Mineral Resources, U.S.S.R.

Feitknechtite, for Walter Feitknecht (1899–   ), Swiss chemist, University of Bern, who first synthesized the compound.

Feldspar (mineral group name), from Swedish names for field (*feldt* or *fält*) and spar (*spat*), in reference to the spar in tilled fields overlying granite; in German the word became *Feldspat*.

Feldspathoid (mineral group name), derived from feldspar (feldspath), in allusion to similarities between the two groups of minerals.

Felsöbanyaite, locality at Felsöbanya, Hungary.

Fenaksite, from composition, containing *Fe, Na, K, Si*.

Ferberite, for Rudolph Ferber of Gera, Germany.

Ferdisilicite, from composition, iron (*ferrum*), two (*di*), and *silicon*.

Fergusonite, for Robert Ferguson (1799–1865), Scottish physician.

Fermorite, for Lewis Leigh Fermor (1880–1954), metallurgist and former director of the Geological Survey of India.

Fernandinite, for Eulagio E. Fernandini, owner of the deposit in Peru where the mineral was discovered.

Feroxyhyte, from composition, iron (*ferric*), *oxygen*, and *hydroxyl*.

Ferrazite, for Jorge Belmiro de Araujo Ferraz, with the Brazilian Geological Survey.

Ferricopiapite, from composition, like copiapite, but with predominant iron (*ferric*).

Ferrierite, for Walter Frederick Ferrier (1865–1950), Canadian geologist and mining engineer.

Ferrihydrite, from composition, iron (*ferric*) and water (*hydro*).

Ferrimolybdite, from composition, iron (*ferric*) and a *molybdate*.

Ferrinatrite, from composition, iron (*ferric*) and sodium (*natrium*).

Ferrisicklerite, from composition, like sicklerite, but with predominant iron (*ferric*).

Ferrisymplesite, from composition, related to symplesite, but with predominant iron (*ferric*).

Ferritungstite, from composition, iron (*ferric*) and a *tungstate*.

Ferroactinolite, from composition, like actinolite, but with predominant iron (*ferrous*).

Ferroaxinite, from composition, like axinite, but with predominant iron (*ferrous*).

Ferrobustamite, from composition, like bustamite, but with predominant iron (*ferrous*).

Ferrocarpholite, from composition, like carpholite, but with predominant iron (*ferrous*).

Ferrocolumbite, from composition, like manganocolumbite, but with predominant iron (*ferrous*).

Ferroedenite, from composition, like edenite, but with predominant iron (*ferrous*).

Ferrohastingsite, from composition, like hastingsite, but with predominant iron (*ferro*us).

Ferrohexahydrite, from composition, iron (*ferro*us) and six (*hexa*) water (*hydro*) molecules.

Ferropumpellyite, from composition, like pumpellyite, but with predominant iron (*ferro*us).

Ferroselite, from composition, iron (*ferr*um) and *sel*enium.

Ferrosilite, from composition, iron (*ferro*us) and *sil*icate.

Ferrotantalite, from composition, like manganotantalite, but with predominant iron (*ferro*us).

Ferrotschermakite, from composition, like tschermakite, but with predominant iron (*ferro*us).

Ferruccite, for Ferruccio Zambonini (1880–1932), Italian mineralogist.

Fersilicite, from composition, iron (*ferr*um) and *silic*on.

Fersmanite, for Alexander Evgenievich Fersman (1883–1945), Russian mineralogist and geochemist, who led expedition to locality in the Khibina Tundras.

Fersmite, for Alexander Evgenievich Fersman (1883–1945), Russian mineralogist and geochemist.

Fervanite, from composition, iron (*ferr*um) and *van*adium.

Fibroferrite, from Latin words for *fiber* and iron (*ferr*um), in allusion to fibrous nature and composition.

Fiedlerite, for Karl Gustav Fiedler (1791–1853), Saxon Commissioner of Mines, who directed an exploratory expedition to Laurium, Greece, where the mineral occurs.

Fillowite, for A. N. Fillow of Branchville, Connecticut, who owned and operated the deposit where mineral was discovered.

Finnemanite, for K. J. Finneman of Langban, Sweden, who first noted the mineral.

Fischesserite, for Raymond Fischesser (1911–), French mineralogist and crystallographer, director of the Ecole National Supérieure des Mines, Paris.

Fizelyite, for Sandor Fizély, mining engineer, who discovered the mineral in Hungary.

Flagstaffite, locality north of Flagstaff, Arizona.

Fleischerite, for Michael Fleischer (1908–), American mineralogist and geochemist, U.S. Geological Survey, Washington, D.C.

Fletcherite, locality at Fletcher mine, Reynolds County, Missouri.

Flinkite, for Gustav Flink (1849–1931), Swedish mineralogist.

Florencite, for William Florence (1864–1942), Brazilian mineralogist who studied the minerals of Minas Geraes, and made a preliminary chemical examination of the mineral.

Flos ferri (var. of aragonite), from Latin for *flower of iron;* not an iron mineral, but associated with iron ore deposits.

Fluellite, from old name for composition, "fluate of alumine," a hydrated aluminum fluoride.

Fluoborite, from composition, a *fluobor*ate of magnesium.

Fluocerite, from composition, a *fluo*ride of *cer*ium and lanthanum.

Fluorapatite, from composition, the *fluor*ine member of the apatite group.

Fluorapophyllite, from composition, the *fluor*ine member of the apophyllite group.

Fluorite, from Latin for *to flow,* since it melts more easily than other minerals with which it was confused.

Foggite, for Forrest F. Fogg, a collector of mineral micromounts from Penacook, New Hampshire.

Formanite, for Francis Gloster Forman, government geologist of Western Australia.

Fornacite, for Lucien Louis Fourneau (1867–1930), former colonial governor of French Congo, Africa, the original locality for the mineral.

Forsterite, for Adolarius Jacob Forster (1739–1806), English mineral collector and dealer, who spent considerable time in St. Petersburg, Russia.

Foshagite, for William Frederick Foshag (1894–1956), American mineralogist, curator at U.S. National Museum, Washington, D.C., who studied minerals at the type locality.

Foshallasite, for William Frederick Foshag (1894–1956), American mineralogist, and part of the mineral name centr*allasite;* gyrolite has replaced the name centrallasite, which originally was named for the fact the type mineral formed a central layer between two other minerals.

Fourmarierite, for Paul Fourmarier (1877–?), Belgian geologist.

Fraipontite, for Julien Jean Joseph Fraipont (1857–1910) and Charles Fraipont, Belgian geologists and paleontologists.

Francevillite, locality in region of Franceville, French Equatorial Africa.

Franckeite, for Carl Francke and Ernest Francke, mining engineers, who were interested in Bolivian geology.

Francoanellite, for Franco Anelli, Italian geographer and speleologist, University of Bari.

Francolite (var. of carbonate-fluorapatite), locality of Wheal Franco at Tavistock, Devonshire, England.

Frankdicksonite, for Frank W. Dickson (1922–), American geochemist, Stanford University.

Franklinite, locality at Franklin, New Jersey.

Franzinite, for Marco Franzini (1938–), Italian mineralogist, University of Pisa.

Freboldite, for Georg Frebold (1891–), German geologist, Hannover.

Freibergite, locality at Freiberg in Saxony.

Freieslebenite, for Johann Karl Freiesleben (1774–1846), a mining commissioner of Saxony.

Fresnoite, locality in Fresno County, California, where mineral was first found.

Freudenbergite, for Wilhelm Freudenberg (1881–), who studied the rocks of Odenwald, Germany, where mineral was found.

Friedelite, for Charles Friedel (1832–1899), French chemist and mineralogist.

Friedrichite, for Otmar Michael Friedrich (1902–), Austrian geologist, Leoben, Styria, who has contributed in the field of ore genesis in Austria.

Fritzscheite, for Carl Julius Fritzsche (1808–1871), German chemist.

Frohbergite, for Max Hans Frohberg (1901–1970), Canadian geologist, Toronto.

Frolovite, locality in Novo-Frolovsk copper deposits, northern Urals, Soviet Union.

Frondelite, for Clifford Frondel (1907–), American mineralogist, Harvard University.

Froodite, locality in Frood mine, Sudbury District, Ontario.

Fuchsite (syn. for chromian muscovite), for Johann Nepomuk von Fuchs (1774–1856), German chemist and mineralogist, Munich.

Fukalite, locality at Fuka, in Okayama Prefecture, Japan.

Fukuchilite, for Nobuyo Fukuchi (1877–1934), Japanese mineralogist and geologist.

Fuloppite, for Béla Fülöpp, Hungarian mineral collector.

Furongite, derived from a poetic description of the Hunan province, China, where the mineral was discovered.

G

Gabrielsonite, for Olof Erik Gabrielson (1912–), Swedish mineralogist, at the Swedish Natural History Museum.

Gadolinite, for Johan Gadolin (1760–1852), Finnish chemist, discoverer of yttrium.

Gagarinite, for Yuri Alekseevich Gagarin (1934–1968), Russian cosmonaut, first man to travel in space.

Gageite, for Robert B. Gage, Trenton, New Jersey, who analyzed the first specimens.

Gahnite, for Johan Gottlieb Gahn (1745–1818), Swedish chemist and mineralogist, the discoverer of manganese.

Gaidonnayite, for Gabrielle (Hamburger) Donnay (1920–), crystallographer and mineralogist, McGill University, Montreal, Canada.

Gaitite, for Robert Irwin Gait (1938–), South African-Canadian mineralogist, Royal Ontario Museum, Toronto.

Galaxite, from the name of the plant galax, abundant in Alleghany County, North Carolina, locality for the mineral.

Galeite, for William Alexander Gale (1898–), American physical chemist, associated with the American Potash and Chemical Corporation in California.

Galena, from Latin *galena* (= *lead ore,* or dross that remains after melting lead).

Galenobismutite, from composition, containing bismuth, and similarity to galena.

Galkhaite, locality at Gal-Khaya, Yakutia, Soviet Union.

Gallite, from composition, containing gallium; first mineral of this element.

Gamagarite, locality at Gamagara ridge, Postmasburg district, South Africa.

Ganomalite, from Greek word meaning *luster,* in allusion to its vitreous to resinous appearance.

Ganophyllite, from Greek for *luster* and *leaf,* in allusion to the high luster on the cleavage laminae.

Garnet (mineral group name), from Latin *granatum* (= *pomegranate*), the seeds of which it was thought to resemble.

Garnierite (general term for hydrous nickel silicates), for Jules Garnier (1839–1904), French engineer, worked in New Caledonia and discovered nickel minerals there.

Garrelsite, for Robert Minard Garrels (1916–), American geologist, Northwestern University, Evanston, Illinois.

Garronite, locality in Garron Plateau area of Antrim, Ireland.

Gaspeite, locality in Gaspé Nord County, Gaspé Peninsula, Quebec, Canada.

Gatumbaite, locality near Gatumba, Gisenyi province, Rwanda.

Gaudefroyite, for Christophe Gaudefroy, French mineralogist who worked in Morocco.

Gaylussite, for Joseph Louis Gay-Lussac (1778–1850), French chemist and physicist.

Gearksutite, from Greek $g\overline{e}$ (= *earth*), in allusion to the earthy aspects of the mineral, and the mineral arksutite (= chiolite), which it was thought to resemble in composition. Arksutite was named from locality near Arksut fjord, South Greenland.

Gedrite, locality in valley of Héas, near Gèdres, France.

Gehlenite, for Adolf Ferdinand Gehlen (1775–1815), German chemist.

Geikielite, for Archibald Geikie (1835–1924), Scottish geologist, director of the Geological Survey.

Genkinite, for A. D. Genkin, Soviet mineralogist who has contributed to the mineralogy and geochemistry of the platinum-group elements.

Genthelvite, from a combination of *Genth* and *helvite,* for Frederick August Ludwig Karl Wilhelm Genth (1820–1893), German-American chemist and mineralogist.

Geocronite, from the Greek for *earth* and *Saturn* (*Kronos*), the alchemistic names for antimony and lead, both contained in the mineral.

Georgiadesite, for Mr. Georgiades, a director of the mine at the locality at Laurium, Greece.

Gerasimovskite, for V. I. Gerasimovskii, Russian mineralogist.

Gerhardtite, for Charles Frederic Gerhardt (1816–1856), chemist who first prepared the artificial compound.

Germanite, from composition, containing germanium.

Gersdorffite, for the von Gersdorffs, owners (ca. 1842) of the nickel mine at Schladming, Styria, Germany.

Gerstleyite, for James Mack Gerstley (1907–   ), English-American industrial executive, Pacific Coast Borax Company, California.

Gerstmannite, for Ewald Gerstmann, mineral collector of Franklin, New Jersey, a collector of minerals from the Franklin and Sterling Hill area.

Getchellite, locality at Getchell mine, Humboldt County, Nevada, at which it was discovered.

Geversite, for Traugott Wilhelm Gevers (1900–), South African geologist.

Gianellaite, for Vincent Paul Gianella (1886–?), American geologist, Mackay School of Mines, University of Nevada.

Gibbsite, for George Gibbs (1776–1833), owner of the mineral collection acquired by Yale early in the nineteenth century.

Giessenite, locality at Giessen, Valais Canton, Switzerland.

Gillespite, for Frank Gillespie, who found the first specimen near his mining claim in Alaska.

Ginorite, for Piero Ginori Conti (1865–1939), of Florence, Italy, a leader in the development of the Tuscan borax industry.

Gismondine, for Carlo Giuseppe Gismondi (1762–1824), Italian mineralogist, Rome.

Gladite, locality at Gladhammar, Sweden.

Glance (obsolete term), from German *glanz* (= *brightness*), in allusion to the bright metallic luster; old term for various ores, e.g., silver glance (argentite), copper glance (chalcocite), lead glance (galena), cobalt glance (cobaltite).

Glauberite, in allusion to the fact it contains considerable quantity of Glauber's salt (Na$_2$SO$_4$·10H$_2$O), named for Johann Rudolf Glauber (1603–1668), German chemist.

Glaucochroite, from Greek for *blue* and *color,* in allusion to its color.

Glaucodot, from Greek for *blue,* because of its use in making smalt, a dark-blue glass.

Glaucokerinite, from Greek for *blue* and *waxlike,* in allusion to its appearance.

Glauconite, from Greek for *blue,* because of its color.

Glaucophane, from Greek for *blue* and to *appear,* because of its color.

Glaukosphaerite, from Greek for *blue* and *sphere,* in allusion to its appearance.

Glockerite, for Ernst Friedrich Glocker (1783–1858), German mineralogist who first described the mineral.

Glucine, from composition, containing beryllium, from glucinum, an old name used for the element.

Gmelinite, for Christian Gottlob Gmelin (1792–1860), German mineralogist and chemist, of Tübingen.

Godlevskite, for M. N. Godlevskii, Russian economic geologist.

Goedkenite, for Virgil Linus Goedken (1940–   ), American chemist and crystallographer, University of Chicago.

Goethite, for Johann Wolfgang von Goethe (1749–1832), German poet and philosopher.

Gold, an Old English word for the metal; perhaps related to the Sanskrit *jval.*

Goldfieldite, locality at Mohawk mine, Goldfield, Nevada.

Goldichite, for Samuel Stephen Goldich (1909–   ), American mineralogist and petrologist, Northern Illinois University, DeKalb.

Goldmanite, for Marcus Isaac Goldman (1881–1965), American sedimentary petrologist, U.S. Geological Survey.

Gonnardite, for Ferdinand Gonnard, French mineralogist, Lyons.

Gonyerite, for Forest A. Gonyer, American analytical chemist, Harvard University.

Gorceixite, for Henrique Gorceix (1842–1919), first director of the School of Mines at Ouro Preto, Brazil.

Gordonite, for Samuel George Gordon (1897–1952), American mineralogist, Academy of Natural Sciences of Philadelphia.

Görgeyite, for Rolf von Görgey (1886–   ), a student of the petrography of Austrian salt deposits.

Goshenite (var. of beryl), locality at Goshen, Massachusetts.

Goslarite, locality at Goslar, in the Harz Mountains, Germany.

Götzenite, for Count Gustav Adolf von Götzen (1886–   ), German traveller, first white man to climb Mt. Shaheru, North Kivu, Belgian Congo, the type locality.

Goudeyite, for Hatfield Goudey, American mining geologist and mineral collector, San Mateo, California, who contributed to the knowledge of the type locality in Nevada.

Gowerite, for Harrison P. Gower, in charge of U.S. Borax and Chemical Corporation mining operations in Death Valley, California.

Goyazite, locality in province of Goyaz, Brazil.

Graemite, Richard Graeme of Ajo, Arizona, a collector and student of Arizona minerals.

Graftonite, locality near Grafton, Grafton County, New Hampshire.

Grandidierite, for Alfred Grandidier (1836–1912), French explorer, who described the geography and natural history of Madagascar.

Grantsite, locality near town of Grants, New Mexico, where mineral was discovered.

Graphite, from Greek word meaning *to write,* in allusion to its use in pencils.

Gratonite, for Louis Caryl Graton (1880–1970), American mining geologist, Harvard University.

Grayite, for Anton Gray, for many years mining adviser to the United Kingdom Atomic Energy Authority.

Greenalite, from English word *green,* in allusion to its color.

Greenockite, for Charles Murray Cathcart (1783–1859) (second earl of Cathcart), whose earlier title was Lord Greenock, the discoverer of the mineral.

Greigite, for Joseph Wilson Greig (1895–1977), American petrologist and physical chemist, Pennsylvania State University.

Grimaldiite, for Frank Saverio Grimaldi (1915–   ), American chemist, U.S. Geological Survey.

Grimselite, locality at Grimsel area, Aar massif, Oberhasli, Bern Canton, Switzerland.

Griphite, from Greek for *puzzle,* in allusion to the complex chemical composition.

Grossular, from Latin *grossularium* (= *gooseberry*), in allusion to the pale green color of some of the specimens.

Groutite, for Frank Fitch Grout (1880–1958), American petrologist, University of Minnesota.

Grovesite, for Arthur William Groves, British geologist, Colonial Geological Surveys.

Grunerite, for Louis Emmanuel Gruner (1809–1883), French chemist of St. Etienne, who first analyzed the mineral.

Grünlingite, for Friedrich Grünling, former curator of the mineral collection, University of Munich.

Guanajuatite, locality at Guanajuato, Mexico.

Guanglinite (= kuanglinite), for Chinese type locality, no specific name given.

Guanine, from its association with guano; guano is derived from Peruvian word *huanu* (= *dung*).

Gudmundite, locality at Gudmundstorp, Sweden.

Guerinite, for Henri Guerin (1906–), French chemist.

Guettardite, for Jean Etienne Guettard (1715–1786), French geologist and mineralogist.

Guildite, for Frank Nelson Guild (1870–1939), University of Arizona.

Guilleminite, for Claude Guillemin, French mineralogist, Bureau de Recherches Géologiques et Minières.

Gummite (general term for secondary uranium oxides), from *gum,* in allusion to the gumlike appearance of some specimens.

Gunningite, for Henry Cecil Gunning (1901–), Canadian geologist, Geological Survey of Canada.

Gustavite, for Gustav Adolf Hageman (1842–1916), chemical engineer, of the Cryolite Firm of Greenland.

Gutsevichite, for V. P. Gutsevich, Soviet geologist of Kazakhstan, Soviet Union.

Guyanaite, locality in Guyana, the country in which the mineral was first found.

Gypsum, from Greek *gypsos* (= *plaster*), an ancient name; was applied equally to gypsum, dehydrated gypsum, lime, and similar materials.

Gyrolite, from Greek for *round,* alluding to its form as crystalline spherical concretions.

H

Haapalaite, for Paavo Haapala, Finnish geologist, Outokumpu Company, Tapiola.

Hackmanite (var. of sodalite), for Victor Axel Hackman (1866–1941), Finnish petrologist.

Hafnon, from composition, a hafnium mineral; in a series with zircon, thus the similar ending.

Hagendorfite, locality at Hagendorf-South pegmatite, Bavaria.

Häggite, for Gunnar Hägg (1903–   ), Swedish chemist, University of Stockholm.

Haidingerite, for Wilhelm Karl von Haidinger (1795–1871), Austrian mineralogist and geologist.

Haiweeite, locality near the Haiwee Reservoir in the Coso Mountains, California.

Hakite, for Jaroslav Hak, Czech mineralogist, Institute of Ore Research, Kutná Hora.

Halite, from Greek word *hals* (= *salt*).

Hallimondite, for Arthur Francis Hallimond (1890–1968), British mineralogist, who contributed to the knowledge of secondary uranium minerals.

Halloysite, for Baron Omalius d'Halloy (1707–1789), who first observed the mineral.

Halotrichite, from Latin *halotrichum*, meaning *salt hair*, in allusion to the silky, fibrous structure it often assumes.

Halurgite, for the Institute of Halurgy, Soviet Union, which has studied saline deposits for many years.

Hambergite, for Axel Hamberg (1863–1933), Swedish mineralogist and geographer, who drew attention to the first specimens.

Hammarite, locality at Gladhammar, Kalmar, Sweden.

Hancockite, for Elwood P. Hancock (1836–1916), Burlington, New Jersey, a collector of minerals from Franklin, New Jersey.

Hanksite, for Henry Garber Hanks (1826–1907), American mineralogist, State Mineralogist of California, in which state the mineral was found.

Hannayite, for James Ballantyne Hannay (1855–?), British chemist, University of Manchester.

Haradaite, for Zyunpei Harada (1898–   ), Japanese mineralogist, professor at Hokkaido University, Hokkaido, who made many contributions to mineralogy.

Hardystonite, locality in Hardyston township, which is the North Hill mine, Franklin, New Jersey.

Harkerite, for Alfred Harker (1859–1939), British petrologist, St. John's College, Cambridge.

Harmotome, from Greek for *joint* and *to cut,* in allusion to the morphology of the twinned crystals.

Harstigite, locality at the Harstig mine, near Pajsberg, Wermland, Sweden.

Hastingsite, locality in Hastings County, Ontario.

Hastite, for P. F. Hast, mining engineer.

Hatchite, for Frederick Henry Hatch (1864–1932), British mining engineer and geologist.

Hatrurite, locality in the Hatrurim formation, Israel.

Hauchecornite, for Wilhelm Hauchecorn (1828–1900), German geologist.

Hauerite, for Joseph Ritter von Hauer (1778–1863) and his son Franz Ritter von Hauer (1822–1899), Austrian geologists.

Hausmannite, for Johann Friedrich Ludwig Hausmann (1782–1859), German mineralogist, University of Göttingen.

Hauyne, for René Just Haüy (1743–1822) French crystallographer and mineralogist.

Hawleyite, for James Edwin Hawley (1897–1965), Canadian mineralogist, Queen's University, Kingston, Ontario.

Haxonite, for H. J. Axon, metallurgist, University of Manchester.

Haycockite, for Maurice Hall Haycock (1900–), head of Mineralogy Section, Department of Energy, Mines and Resources, Ottawa, Canada.

Heazlewoodite, locality at Heazlewood, Tasmania.

Hectorite, locality at Hector, California.

Hedenbergite, for M. A. Ludwig Hedenberg, Swedish chemist, who analyzed and described the mineral.

Hedleyite, locality near Hedley, Osoyoos mining district, British Columbia.

Hedyphane, from Greek for *beautifully bright,* in allusion to the high luster.

Heideite, for Fritz Heide (1891–1973), professor at Jena, Germany, who contributed much to the field of meteoritics.

Heidornite, for F. Heidorn, geologist.

Heinrichite, for Eberhardt William Heinrich (1918–), American mineralogist, University of Michigan, Ann Arbor.

Heliodor (var. of beryl), from Greek for *sun* and *gilded,* in allusion to its yellow color.

Heliophyllite, from Greek for *sun* and *leaf,* in allusion to the yellow color and foliated structure.

Heliotrope (var. of quartz), from Greek for *sun* and *turning*, in allusion to the red appearance obtained when translucent green chalcedony with red spots is immersed in water in sunlight.

Hellandite, for Amund Theodor Helland (1846–1918), Norwegian geologist of Oslo.

Hellyerite, for Henry Hellyer, who conducted surveys and exploration in northwestern Tasmania in the period 1826–1830.

Helvite, from Greek for *sun*, in allusion to its yellow color.

Hemafibrite, from Greek for *blood* and *fiber*, in allusion to its color and structure.

Hematite, from Greek *haimatitis* (= *blood red*), in allusion to its color.

Hematolite, from Greek for *blood*, in allusion to its color.

Hematophanite, from Greek for *blood* and *visible*, apparently in allusion to the blood-red color in transmitted light.

Hemihedrite, from morphology of the crystals which exhibit triclinic *hemihedral* symmetry.

Hemimorphite, from the *hemimorphic* symmetry of the crystals.

Hemusite, from Hémus, the ancient name of the Balkan Mountains where the mineral occurs.

Hendersonite, for Edward Porter Henderson (1898–), student of meteorites, U.S. National Museum, Washington, D.C.

Hendricksite, for Sterling B. Hendricks (1902–), American crystallographer and chemist, a student of micas.

Henritermierite, for Henri F. E. Termier (1897–), French geologist, professor at the Sorbonne, Paris.

Hercynite, from Latin name of the Bohemian forest, *Silva Hercynia*, where the mineral was first found.

Herderite, for Siegmund August Wolfgang von Herder (1776–1838), mining official at Freiberg, Saxony.

Herschelite, for John Frederick William Herschel (1792–1871), British astronomer.

Herzenbergite, for Robert Herzenberg (1885–), German chemist.

Hessite, for Germain Henri Hess (1802–1850), Swiss-Russian chemist, St. Petersburg, Russia.

Hessonite (var. of grossular), from Greek for *inferior*, in reference to its hardness.

Hetaerolite, from Greek word for *companion*, in allusion to its association with chalcophanite in some specimens.

Heterogenite, from Greek for *other* and *kind*, in reference to its difference in composition from certain minerals it otherwise resembles.

Heteromorphite, from Greek for *other* and *form*, in allusion to the difference in form between this species and a supposed dimorphous mineral.

Heterosite, from Greek for *another*, probably because it was the second manganese-containing mineral from the same locality.

Heubachite (var. of heterogenite), locality in the Heubachthal, near Wittichen, Baden.

Heulandite, for John Henry Heuland (1778–1856), English mineral collector and dealer.

Hewettite, for Donnel Foster Hewett (1881–1971), American geologist, U.S. Geological Survey.

Hexagonite (var. of tremolite), from its supposed hexagonal structure, later shown to be monoclinic.

Hexahydrite, from composition, in allusion to its six (*hexa*) waters (*hydro*).

Hexahydroborite, from composition, a calcium *bor*ate with six (*hexa*) waters (*hydro*).

Hexatestibiopanickelite, from *hexa*gonal system and its composition, containing *Te*, Sb (*stibi*um), Pd (*pa*lladium), and *nickel*.

Heyite, for Max Hutchinson Hey (1904–), British chemist and mineralogist, British Museum (Natural History), London.

Heyrovskyite, for Jaroslav Heyrovský (1890–1967), Czech physiochemist, Nobel Laureate in chemistry.

Hibonite, for Mr. P. Hibon, who discovered the mineral.

Hidalgoite, locality in the state of Hidalgo, Mexico.

Hiddenite (var. of spodumene), for William Earl Hidden (1853–1918), American mineralogist, New York.

Hieratite, locality at the crater of Vulcano (Greek name, *Hiera*), on one of the Lipari Islands off Sicily.

Hilairite, locality at Mont St. Hilaire, Quebec, Canada.

Hilgardite, for Eugene Woldemar Hilgard (1833–1916), German-American geologist, who described saline deposits of Louisiana.

Hillebrandite, for William Francis Hillebrand (1853–1925), American geochemist.

Hinsdalite, locality in Hinsdale County, Colorado, where the mineral was discovered.

Hiortdahlite, for Thorstein Hallager Hiortdahl (1839–1925), Norwegian mineralogist, of Christiania.

Hisingerite, for Wilhelm Hisinger (1766–1852), Swedish chemist and mineralogist.

Hjelmite, for Peter Jacob Hjelm (1746–1813), Swedish chemist.

Hocartite, for Raymond Jean Hocart (1896–), French crystallographer, University of Paris.

Hodgkinsonite, for H. H. Hodgkinson, who discovered the mineral.

Hodrushite, locality at Banská-Hodrusa near Banská-Stiavnica, Czechoslovakia.

Hoelite, for Adolf Hoel (1879–?), Norwegian geologist.

Hoernesite, for Moritz Hoernes (1815–1868), curator of the Imperial Mineral Cabinet in Vienna.

Högbomite, for Arvid Gustaf Högbom (1857–1940), Swedish geologist, University of Uppsala.

Hohmannite, for Thomas Hohmann, mining engineer of Valparaiso, Chile, who discovered the mineral.

Holdenite, for Albert Fairchild Holden (1866–1913), mining engineer and mineral collector, in whose collection the mineral was first noticed.

Hollandite, for Thomas Henry Holland (1868–1947), British geologist, Director of the Geological Survey of India.

Hollingworthite, for Sydney Ewart Hollingworth (1899–1966), British geologist, University College, London.

Holmquistite, for Per Johan Holmquist (1866–1946), Swedish petrologist, Stockholm.

Holtite, for Harold Edward Holt (1908–1967), Prime Minister of Australia from 1966 to 1967.

Homilite, from Greek for *to occur together*, in allusion to its association with meliphanite.

Honessite, for Arthur Pharaoh Honess (1886–1942), American mineralogist, Pennsylvania State University.

Hongquiite (= hungchiite), for Chinese type locality, no specific name given.

Hongshiite (= hungshihite), locality in the Hung district of China.

Hopeite, for Thomas Charles Hope (1766–1844), Scottish chemist, University of Edinburgh.

Hornblende, from German *Horn* (= *horn*) and *blenden* (= *to blind* or *deceive*); and old German word for any dark prismatic mineral occurring in ores but containing no recoverable metal.

Horsfordite, for Eben Norton Horsford (1818–1893), American chemist, Harvard University.

Howieite, for Robert Andrew Howie (1923–), English mineralogist, King's College, London University.

Howlite, for Henry How (1828–1879), Canadian chemist, University of King's College, Windsor, Nova Scotia, who first reported the mineral.

Hsianghualite, from Chinese for *fragrant flower.*

Huanghoite, locality near the Huang Ho River in China.

Huebnerite, for Adolph Hübner, German mining engineer, of Freiberg, Saxony.

Huemulite, locality at Huemul mine, Mendoza Province, Argentina.

Hügelite, for F. Hügel.

Hühnerkobelite, locality at Hühnerkobel, Bavaria.

Hulsite, for Alfred Hulse Brooks (1871–1924), American geologist, U.S. Geological Survey.

Humberstonite, for James Thomas Humberstone (1850–1939), chemist, who worked with saline minerals of Chilean nitrate deposits.

Humboldtine, for Baron Alexander von Humboldt (1769–1859), German naturalist.

Humite, for Abraham Hume (1749–1838), English connoisseur and collector of works of art, precious stones, and minerals.

Hummerite, locality at Hummer mine, Montrose County, Colorado.

Hungchaoite, for Chang Hung-chao (1877–1951), Chinese geologist and mineralogist.

Huntite, for Walter Frederick Hunt (1882–1975), American mineralogist, University of Michigan, Ann Arbor.

Hureaulite, locality at Huréaux in St. Sylvestre, Haute Vienne, France.

Hurlbutite, for Cornelius Searle Hurlbut, Jr. (1906–), American mineralogist, Harvard University.

Hutchinsonite, for Arthur Hutchinson (1866–1937), English mineralogist, Cambridge.

Huttonite, for Colin Osborne Hutton (1910–1971), New Zealand-American mineralogist, Stanford University.

Hyacinth (var. of zircon), named by the ancients for its similarity in color to purplish blue hyacinth flowers; originally may have applied to sapphire.

Hyalite (var. of opal), from Greek for *glass,* in allusion to its appearance.

Hyalophane, from Greek for *glass* and *to appear,* in allusion to its transparent crystals.

Hyalotekite, from Greek for *glass* and *to melt,* because it fuses to a clear glass in a blowpipe flame.

Hydroastrophyllite, from its water (*hydro*) content and similarity to astrophyllite in composition.

Hydrobasaluminite, from its water (*hydro*) content and similarity to basaluminite in composition.

Hydroboracite, from its water (*hydro*) content and similarity to boracite in composition.

Hydrocalumite, from composition, water (*hydro*), *calc*ium, and *alum*inum.

Hydrocerussite, from its water (*hydro*) content and similarity to cerussite in composition.

Hydrochlorborite, from composition, a *bor*ate with water (*hydro*) and *chlor*ine.

Hydrodresserite, from its water (*hydro*) content, and its chemical relationship to dresserite, which it forms when dehydrated.

Hydrogarnet (mineral group name), from its water (*hydro*) content; a garnet in which SiO_4 is partly replaced by $(OH)_4$.

Hydroglauberite, from its water (*hydro*) content and similarity to glauberite in composition.

Hydrogrossular, from its water (*hydro*) content; grossular in which SiO_4 is partly replaced by $(OH)_4$.

Hydrohalite, from its water (*hydro*) content and similarity to halite in composition.

Hydrohetaerolite, from its water (*hydro*) content and similarity to hetaerolite in composition.

Hydromagnesite, from its water (*hydro*) content and similarity to magnesite in composition.

Hydromica (mineral group name), from water (*hydro*) and mica, in allusion to the fact that water can substitute for some of the usual ions in the structure.

Hydromolysite, from its water (*hydro*) content and similarity to molysite in composition.

Hydronium jarosite, from composition, like jarosite, but with hydronium ion $(H_3O)^+$.

Hydrophilite, from Greek words for *water* (*hydro*) and *friend,* in allusion to its hygroscopic properties.

Hydroromarchite, from its water (*hydro*) content and similarity to romarchite in composition.

Hydroscarbroite, from its water (*hydro*) content and similarity to scarbroite in composition.

Hydrotalcite, from its water (*hydro*) content and similarity to talc in physical properties and composition.

Hydrotroilite, from its water (*hydro*) content and similarity to troilite in composition.

Hydrotungstite, from its water (*hydro*) content and similarity to tungstite in composition.

Hydrougrandite, from its water (*hydro*) content; ugrandite in which SiO_4 is partly replaced by $(OH)_4$.

Hydroxyapophyllite, from composition, the *hydroxy*al member of the apophyllite group.

Hydroxyl-apatite, from composition, the *hydroxyl* member of the apatite group.

Hydroxyl-bastnaesite, from composition, like bastnaesite, but with the hydroxyl ion.

Hydroxyl-ellestadite, from composition, like ellestadite, but with the hydroxyl ion.

Hydroxyl-herderite, from composition, like herderite, but with the hydroxyl ion.

Hydrozincite, from composition, water (*hydro*), *zinc*, and others.

Hypersthene, from Greek for *very strong*, because its hardness seemed greater than other minerals (hornblende) with which it was originally confused.

I

Ianthinite, from Greek for *violet colored*, in allusion to the color of the mineral.

Ice, from Middle English *is, iis*, derived from Old English *is;* related to Dutch *ijs* and German *Eis*.

Idaite, locality at Ida mine, Khan, South West Africa.

Idocrase (syn. for vesuvianite), from Greek for *form* and *mixture*, because the crystal forms present appear to be a combination of those found on other minerals.

Idrialite, locality at Idria, Italy.

Iimoriite, for Satoyasu Iimori (1885–   ) and Takeo Iimori, of Japan, who described many rare-earth minerals.

Ikaite, locality at Ika fjord, near Ivigtut, Greenland.

Ikunolite, locality at Ikuno mine, Hyogo Prefecture, Japan.

Ilesite, for Malvern Wells Iles (1852–1890), American metallurgist, of Denver, Colorado.

Ilimaussite, locality at Ilimaussaq, South Greenland.

Illite (mineral group name), locality in the state of Illinois, source of many specimens which have been studied.

Ilmajokite, locality at Ilmajok River, Lovozero Tundra, Kola Peninsula, Soviet Union.

Ilmenite, locality in the Ilmen Mountains, Soviet Union, where it was found.

Ilmenorutile, locality in the neighborhood of Miask, Ilmen Mountains, Soviet Union, and similarity to rutile.

Ilsemannite, for J. C. Ilsemann (1727–1822), a Mining Commissioner at Clausthal, Harz, Germany.

Ilvaite, locality on the island of Elba (Latin, *Ilva*).

Imgreite, for the Institute of the Mineralogy, Geochemistry, and Crystal Chemistry of Rare Elements (IMGRE), Soviet Union.

Imhofite, for Josef Imhof, professional mineral collector, Binn, Switzerland.

Incaite, for the Incas, the first recorded miners of the type locality in Bolivia.

Inderborite, locality at Inder Lake, western Kazakhstan, Soviet Union, and composition, a borate.

Inderite, locality near Inder Lake, western Kazakhstan, Soviet Union.

Indialite, locality in India, the country in which the mineral was first found.

Indicolite (var. of elbaite), from Latin *indicum* (= *indigo*), in allusion to its color.

Indigirite, locality near Indigirki River, northeastern Yakutia, Soviet Union.

Indite, from composition, containing indium.

Indium, from brilliant indigo-colored line in its spectrum.

Inesite, from Greek word for *flesh fibers,* because it is found in fibrous masses of flesh-red color.

Innelite, from Inneli, the Yakut name for the Inagli River, southern Yakutia, Soviet Union, the locality of the mineral.

Insizwaite, locality in the Insizwa deposit, South Africa.

Inyoite, locality in Inyo County, California.

Iodargyrite, from composition, involving *iod*ine and silver (Greek, *argyros*).

Iodobromite (syn. for iodian bromargyrite), from composition, including *iod*ine, *brom*ine and other elements.

Iowaite, locality in the state of Iowa.

Iranite, locality in Iran, the country in which the mineral was first found.

Iraqite, locality in Iraq, the country in which the mineral was first found.

Irarsite, from composition, it contains *ir*idium, *ar*senic, and *s*ulfur.

Irhtemite, locality in the Irhtem ore deposit, Morocco.

Iridarsenite, from composition, *irid*ium and *arsen*ic.

Iridium, from Latin *iris* (= *rainbow*), because salts of the element are highly colored.

Iridosmine, from composition, *irid*ium and *osm*ium.

Iron, an Old English word for the metal.

Ishikawaite, locality in Ishikawa district, Iwaki Province, Japan.

Isoclasite, from Greek words for *equal* and for *fracture,* because it cleaves as easily as gypsum.

Isoferroplatinum, for crystal structure (*iso*metric) and composition, iron (*ferr*um) and *platinum.*

Isokite, locality near Isoka, Northern Rhodesia.

Isomertieite, from Greek word for *equal* and mineral mertieite, in allusion to their dimorphous relationship.

Isoplatinocopper, from crystal system, *iso*metric, and from composition, *platin*um and *copper.*

Itoite, for Tei-ichi Ito (1898–), Japanese mineralogist, University of Tokyo.

Ixiolite, for Ixion, a mythological person related to Tantalus; in allusion to the relationship of the mineral to tantalite (named for Tantalus).

J

Jacobsite, locality at Jacobsberg, Wermland, Sweden.

Jade (general term for gem nephrite or jadeite), from Spanish term *piedra de yjada* (= *stone of the side*); so called because the stone was supposed to cure side pains.

Jadeite, name derived from jade, because many jade specimens were shown to be composed of the mineral.

Jagoite, for John B. Jago, American mineral collector of San Francisco, California.

Jagowerite, for John Arthur Gower (1921–1972), Canadian mineralogist, University of British Columbia.

Jahnsite, for Richard Henry Jahns (1915–), American pegmatite scholar, Dean of Earth Sciences, Stanford University.

Jalpaite, locality at Jalpa, Mexico.

Jamborite, for John Leslie Jambor (1936–), Canadian mineralogist, Geological Survey of Canada, Ottawa.

Jamesonite, for Robert Jameson (1774–1854), Scottish mineralogist of Edinburgh.

Janggunite, locality in the Janggun mine, Bonghwa, Korea.

Jarlite, for Carl Frederik Jarl (1872–?), official in the Danish cryolite industry.

Jarosite, locality in the Jaroso ravine in the Sierra Almagrera, Spain.

Jasper (var. of quartz), derived from Latin *iaspis,* which is of oriental origin, corresponding to the Persian *iashm* and *jashp* and the Assyrian *ashpu.*

Jefferisite (var. of vermiculite), for William Walter Jefferis (1820–1906), noted mineral collector of West Chester, Pennsylvania.

Jeffersonite (var. of augite or aegirine), for Thomas Jefferson (1743–1826), American statesman and third President.

Jennite, for Clarence M. Jenni (1896–   ), American mineral collector, director of the geological museum at the University of Missouri.

Jeremejevite, for Pavel Vladimirovitch Jeremejev (1830–1899), Russian mineralogist and engineer.

Jeromite, locality at the United Verde mine, Jerome, Arizona.

Jimboite, for Kotora Jimbo, Japanese mineralogist, founder of the Mineralogical Institute, University of Tokyo.

Jimthompsonite, for James Burleigh Thompson, Jr. (1921–   ), American petrologist and geochemist, Harvard University, who predicted the existence of the mineral.

Joaquinite, locality at the Joaquin ridge of the Diablo Mountain range, California.

Joesmithite, for Joseph Victor Smith (1928–   ), English-American mineralogist, University of Chicago.

Johachidolite, locality in Johachido district, Kenkyohokudo Prefecture, North Korea.

Johannite, for Johann Baptist Joseph Fabian Sebastian (1782–1859), Archduke of Austria and founder of the Styrian Landesmuseum in Graz.

Johannsenite, for Albert Johannsen (1871–1962), American geologist-petrologist, University of Chicago.

Jokokuite, locality at the Jokoku manganese-lead-zinc mine, southwestern Hokkaido, Japan.

Joliotite, for Frederic Joliot (1900–1958), French physicist.

Jonesite, for Francis T. Jones, American mineral collector, Berkeley, California, who first noticed the mineral in 1957.

Jordanite, for Dr. J. Jordan of Saarbrücken, in the Saar Basin, who furnished specimens for the original study.

Jordisite, for Eduard Friedrich Alexander Jordis (1868–1917), colloidal chemist.

Joseite, locality at São José, Brazil.

Jouravskite, for Georges Jouravsky (1896–1964), chief geologist of the Division de la Géologie du Maroc (Morocco).

Juanite, locality in San Juan region, Colorado.

Julgoldite, for Julian Royce Goldsmith (1918–   ), American geochemist, University of Chicago.

Julienite, for Henri Julien (?–1920), who discovered the mineral in Katanga, Belgian Congo.

Junitoite, for Jun Ito (1926–1978), Japanese-American mineralogist and chemist, University of Chicago, who first noted the compound in synthetic silicates.

Junoite, locality at Juno ore body, Tennant Creek, Australia.

Jurbanite, for Joseph John Urban (1915–), American mineral collector of Tucson, Arizona, who first noticed the mineral.

K

Kaersutite (var. of hornblende), locality at Kaersut, northern Greenland.

Kafehydrocyanite, from composition, potassium (*ka*lium), iron (*fe*rrum), water (*hydro*), and *cyan*ide.

Kahlerite, for Franz Kahler, Austrian geologist, Carinthian Landesmuseum, Klagenfurt.

Kainite, from Greek word for *recent,* alluding to its recent (secondary) formation.

Kainosite, from Greek word for *unusual,* in allusion to the composition, a silicate with carbonate.

Kaliborite, from composition, potassium (*kali*um) and *bor*ate.

Kalicinite, from composition, it contains potassium (*kali*um).

Kalinite, from composition, it contains potassium (*kali*um).

Kaliophilite, from composition, potassium (*kali*um), and Greek for *friend,* in allusion to the presence of the element.

Kalipyrochlore, from composition, like pyrochlore, but with predominant potassium (*kali*um).

Kalistrontite, from composition, potassium (*kali*um) and *stront*ium.

Kalsilite, from composition, *K, Al, Si;* formula is $KAlSiO_4$.

Kamacite, from Greek for *pole* or *shaft,* in allusion to platy structure.

Kämmererite (var. of clinochlore), for A. A. Kämmerer (1789–1858), Russian mining director of St. Petersburg.

Kandite (mineral group name), from the minerals *ka*olinite, *n*acrite, and *d*ickite.

Kanemite, locality at Kanem region, northeastern edge of Lake Chad, Africa.

Kankite, locality at the very old dumps near Kank, Kutná Hora district, Czechoslovakia.

Kanoite, for Hiroshi Kano, Japanese petrologist, Akita University.

Kaolinite, from kaolin, a corruption of the Chinese *káuling* (= *high ridge*), the name of a hill near Jauchu Fa, where the mineral occurs.

Karelianite, locality in Karelian schist belt of Finland.

Karibibite, locality at Karibib pegmatite, South West Africa.

Karnasurtite, locality at Karnasurt Mountain, Lovozero alkaline massif, Soviet Union.

Karpatite, apparently for locality in Trans-Carpathians.

Karpinskite, for Alexander Petrovich Karpinski (1846–1936), Russian geologist, late President of the Academy of Sciences, Soviet Union.

Kasoite (var. of celsian), locality at Kaso mine, Japan.

Kasolite, locality at Kasolo, Katanga district, Belgian Congo.

Kassite, for Nikolai Grigorevich Kassin, Russian geologist, discoverer of the Afrikanda massif, Kola Peninsula.

Katophorite, from Greek for *carrying down,* in allusion to its volcanic origin.

Katoptrite, from Greek for *mirror,* in allusion to its luster.

Kawazulite, locality at Kawazu mine, Shizuoka Prefecture, Japan.

Kazakovite, for M. E. Kazakova, Russian chemist who analyzed the mineral.

Keatite, for Paul Powell Keat (1929–   ), student of ceramics, Norton Company, Worcester, Massachusetts, who discovered the synthetic phase.

Kegelite, for Friedrich Wilhelm Kegel, director (1922–1938) of the Tsumeb mine, South West Africa.

Kehoeite, for Henry Kehoe who first observed the mineral at Galena, South Dakota.

Keldyshite, for Mstislav Vsevolodovich Keldysh (1911–   ), Russian mathematician and former President of the Academy of Sciences, Soviet Union.

Kellyite, for William Crowley Kelly (1929–   ), American geologist, University of Michigan, Ann Arbor.

Kemmlitzite, locality at Kemmlitz deposit, Saxony.

Kempite, for James Furman Kemp (1859–1926), American geologist, Columbia University, New York.

Kennedyite, for William Quarrier Kennedy, British geologist, University of Leeds.

Kentrolite, from Greek for *spike,* in allusion to its prismatic habit.

Kenyaite, locality at Lake Magada, Kenya, Africa.

Kermesite, from *kermes,* a name given in old chemistry to red amorphous antimony trisulfide often mixed with antimony trioxide; named in allusion to its composition.

Kernite, locality in Kern County, California.

Kerolite (var. of talc with disordered structure), from Greek word for *wax*, in allusion to the waxy appearance of the material.

Kesterite, locality at Kester deposit, Yano-Àdychansk region, Soviet Union.

Kettnerite, for Radim Kettner (1891–), Czech geologist, Charles University, Prague.

Keyite, for Charles L. Key, American mineral dealer from Canton, Connecticut, who supplied the mineral for investigation.

Khademite, for N. Khadem, Director of the Geological Survey of Iran.

Khibinskite, locality in Gakman Valley, Khibina massif, Kola Peninsula, Soviet Union.

Kidwellite, for Albert Laws Kidwell (1919–), American geologist, Exxon Production Research Company, Houston, Texas, who collected the original specimens.

Kieserite, for Dietrich Georg Kieser (1779–1862), President of the Jena Academy, Jena, Germany.

Kilchoanite, locality at Kilchoan, Ardnamurchan, Scotland.

Killalaite, locality near Killala Bay, Ireland.

Kimzeyite, for the Kimzey family, long known in connection with the mineralogy of Magnet Cove, Arkansas; mineral was discovered by Joe Kimzey.

Kingite, for D. King, geologist, Department of Mines, South Australia, who first collected the mineral.

Kinoite, for Eusebio Francisco Kino (1645–1711), celebrated Jesuit explorer of the southwestern United States.

Kinoshitalite, for Kameki Kinoshita (1896–1974), Japanese geologist, Kyushu University, investigator of ore deposits of Japan.

Kirschsteinite, for Egon Kirschstein, German geologist, pioneer in geological work in North Kivu, Zaire.

Kitkaite, locality in valley of the Kitka River in Kuusamo, northeast Finland.

Kladnoite, locality in coal basin of Kladno, Bohemia.

Klebelsbergite, for Kuno Klebelsberg (1875–1932), Hungarian educator.

Kleinite, for Carl Klein (1842–1907), German mineralogist, University of Berlin.

Klockmannite, for Friedrich Klockmann (1858–1937), German mineralogist, Technischen Hochschule, Aachen.

Knipovichite (var. of alumohydrocalcite), for Yuliya Nikolaevna Knipovich, Soviet mineralogist, who participated in the study of the mineral.

Knorringite, for Oleg von Knorring, Department of Earth Sciences, Leeds University, Leeds, England.

Koashvite, locality at Mt. Koashva, in the Khibina massif, Kola Peninsula, Soviet Union.

Kobeite, locality near Kobe-mura, Nakagun, Kyoto Prefecture, Japan.

Kobellite, for Wolfgang Franz von Kobell (1803–1882), German mineralogist.

Koechlinite, for Rudolf Ignatz Koechlin (1862–1939), Austrian mineralogist, curator at Hof-Museum, Vienna.

Koenenite, for Adolph von Koenen (1837–1915), German geologist of Göttingen, who first found the mineral.

Kogarkoite, for Lia N. Kogarko, Russian geochemist who first noticed the occurrence of the mineral in the Soviet Union.

Koktaite, apparently for Jaroslav Kokta, Czech chemist, who analyzed the artificial compound.

Kolbeckite, for Friedrich Ludwig Wilhelm Kolbeck (1860–1943), German mineralogist, Mining Academy, Freiberg.

Kolovratite, for L. S. Kolovrat-Chervinsky, Russian radiologist.

Komarovite, for Vladimir M. Komarov (1927–1967), Russian cosmonaut, who was killed during his return from a space flight on April 23, 1967.

Koninckite, for Laurent Guillaume de Koninck (1809–1887), Belgian geologist.

Kornelite, for Kornel Hlavacsek, Hungarian.

Kornerupine, for Andreas Nikolaus Kornerup (1857–1881), Danish geologist.

Korzhinskite, for Dmitry Sergeevich Korzhinsky (1899–), Russian geologist and petrographer, Academician of the Academy of Sciences, Soviet Union.

Kostovite, for Ivan Kostov (1913–), Bulgarian mineralogist, Sofia.

Kotoite, for Bundjiro Koto (1856–1935), Japanese geologist and petrographer, University of Tokyo.

Köttigite, for Otto Köttig (1824–?), chemist of Schneeberg, Saxony.

Kotulskite, for Vladimir Klementevich Kotulskii, Russian geologist, a student of copper–nickel sulfide deposits.

Koutekite, for Jaromir Koutek (1902–), Czech economic geologist, Charles University, Prague.

Kozulite, for Shukusuke Kozu (1880–1955), Japanese mineralogist, Tohoku University, who contributed to the study of rock-forming minerals.

Kraisslite, for Frederick Kraissl, Jr. (1899–), and his wife Alice L.

(Plenty) Kraissl (1905–   ), American amateur mineralogists of Hackensack, New Jersey.

Kratochvilite, for Josef Kratochvil (1878–?), Czech petrographer.

Krausite, for Edward Henry Kraus (1875–1973), American mineralogist, University of Michigan, Ann Arbor.

Krauskopfite, for Konrad Bates Krauskopf (1910–   ), American geologist and geochemist, Stanford University.

Krautite, for François Kraut, French mineralogist, Museum national d'Histoire naturelle, Paris.

Kremersite, for Peter Kremers (1827–?), German chemist.

Krennerite, for Joseph A. Krenner (1839–1920), Hungarian mineralogist.

Kribergite, locality, probably contraction of Kristineberg mine, Västerbotten Province, northern Sweden.

Krinovite, for Evgeny Leonidovich Krinov, Russian student of meteorites.

Kröhnkite, for Mr. B. Kröhnke, who first analyzed the mineral.

Krupkaite, locality at Krupka, Czechoslovakia.

Krutaite, for Thomas Kruta, Czech mineralogist, Moravian Museum, Brno.

Krutovite, for Georgi Alekseevich Krutov, Soviet mineralogist, Moscow State University, who has studied nickel–cobalt deposits.

Kruzhanovskite, for Vladimir Ilitch Kruzhanovsky, Russian mineralogist, curator of Mineralogical Museum, Academy of Sciences, Soviet Union.

Ktenasite, for Konstantin Anton Ktenas (1884–1935), Greek mineralogist and geologist.

Kulanite, for Alan Kulan (1921–1977), Ross River, Yukon Territory, Canada, who submitted the original specimens for study.

Kullerudite, for Gunnar Kullerud (1921–   ), Norwegian-American geochemist, Purdue University, West Lafayette, Indiana.

Kunzite (var. of spodumene), for George Frederick Kunz (1856–1932), American mineralogist and gemologist.

Kupletskite, for Boris Mikhailovich Kupletski (1894–?), and Elsa Maximilianovna Bohnshtedt–Kupletskaya (?–1974), Russian geologists.

Kuranakhite, locality in Kuranakh gold deposit, southern Yakutia, Soviet Union.

Kurchatovite, for Igor Vasilievich Kurchatov (1903–1960), Russian nuclear physicist.

Kurgantaite, locality in western Kurgan-Tau (western Kazakhstan), Soviet Union.

Kurnakovite, for Nikolai Semenovich Kurnakov (1860–1941), Russian mineralogist and chemist.

Kurumsakite, locality at Kara-Tau, Soviet Union; Russian name not given.

Kusuite, locality at Kusu vanadium deposit, southwest of Kinshasa, in lower Zaire.

Kutinaite, for J. Kutina, Lecturer at the Charles University, Prague, Czechoslovakia.

Kutnohorite, locality at Kutnahora (Kutná Hora), Bohemia.

Kyanite, from Greek word for *blue*, in allusion to its most common color.

L

Labradorite, locality in Labrador; first brought from the Isle of Paul off Labrador in about 1770.

Labuntsovite, for Aleksander Nikolaevich Labuntsov and Ekaterina Eutikhieva Labuntsov-Kostyleva, Russian mineralogists.

Lacroixite, for François Antoine Alfred Lacroix (1863–1948), French mineralogist.

Laffittite, for Pierre Laffitte (1925–   ), French mathematical geologist, Paris.

Laihunite, locality near Lai-He village, northeastern China.

Laitakarite, for Aarne Laitakari (1890–   ), Finnish geologist, Geological Survey of Finland.

Lamprophyllite, from Greek for *shining* and *leaf*, in allusion to its luster and cleavage.

Lanarkite, locality in Lanarkshire, Scotland.

Landauite, for Lev Davidovich Landau (1908–1968), Soviet physicist.

Landesite, for Kenneth Knight Landes (1899–   ), American geologist, University of Michigan, Ann Arbor, a student of pegmatites.

Langbanite, locality at Långban, Wermland, Sweden.

Langbeinite, for A. Langbein, German chemist of Leopoldshall.

Langisite, locality at Langis mine, Casey township, Ontario.

Langite, for Victor von Lang (1838–1921), Austrian physicist and physical crystallographer, University of Vienna.

Lansfordite, locality near Lansford, Carbon County, Pennsylvania.

Lanthanite, from composition, it contains lanthanum.

Lapis lazuli (var. of lazurite), from Latin for *stone* and Persian *lazhward*, meaning blue color.

Laplandite, locality in the area called Lapland, at Karnasurt Mountain, Lovozero alkalic massif, Kola Peninsula, Soviet Union.

Larderellite, for Francesco de Larderell (1848–1925), a proprietor of the Tuscan borax industry, Italy.

Larnite, locality at Larne in County Antrim, Northern Ireland.

Larosite, for Fred LaRose, Canadian blacksmith, one of the discoverers of silver at Cobalt, Ontario.

Larsenite, for Esper Signius Larsen, Jr. (1879–1961), American petrologist, Harvard University.

Latiumite, locality at Albano, Latium, Italy.

Latrappite, locality near the community of La Trappe, Quebec.

Laubmannite, for Heinrich Laubmann (1865–1951), German mineralogist.

Laueite, for Max von Laue (1879–1960), German physicist, first to use crystals as diffraction gratings, thus providing technique for the study of atomic structures of crystals.

Laumontite, for François Pierre Nicolas Gillet de Laumont (1747–1834), French, who first found the mineral in Brittany.

Launayite, for Louis A. A. de Launay (1860–1938), French geologist, who studied the origin of mineral deposits.

Laurionite, locality at Laurium (Laurion), Greece.

Laurite, for Laura R. Joy, wife of Charles Arad Joy (1823–1891), American chemist, Columbia University, New York.

Lausenite, for Carl Lausen, mining engineer, who first described the species.

Lautarite, locality at Oficina Lautaro, Antofagasta Province, Chile.

Lautite, locality at Lauta, near Marienberg, Saxony.

Lavendulan, from Latin word *lavendula* (= *lavender*), in allusion to the color of the mineral.

Lavenite, locality on island of Låven, Langesund fjord, southern Norway.

Lawrencite, for John Lawrence Smith (1818–1883), American chemist, mineralogist, and student of meteorites.

Lawsonite, for Andrew Cowper Lawson (1861–1952), Scottish-American geologist, University of California.

Lazulite, from Persian *lazhward* (= *blue*).

Lazurite, from Persian *lazhward* (= *blue*).

Lead, an Old English word for the metal; perhaps derived from Celtic; related to Dutch *lood* and German *Lot*.

Leadhillite, locality at Leadhills, Lanarkshire, Scotland.

Lechatelierite, for Henri Le Chatelier (1850–1936), French chemist and metallurgist.

Lecontite, for John Lawrence LeConte (1825–1883), American entomologist of Philadelphia, who discovered the mineral.

Legrandite, for Mr. Legrand, Belgian mine manager, who collected the first specimen.

Lehiite, locality near Lehi, Utah County, Utah.

Leifite, for Leif Ericson (10th century), Norse mariner and adventurer.

Leightonite, for Tomas Leighton, mineralogist, University of Santiago, Chile.

Leiteite, for Luis Teixeira-Leite, mineralogist, Pretoria, Republic of South Africa, the discoverer.

Lemoynite, for Charles Lemoyne (1625–1685) and his four sons, prominent in the history of Canada.

Lengenbachite, locality at Lengenbach, in the Binnental, Valais, Switzerland.

Lenoblite, for André Lenoble, French mineralogist-geologist, who studied geology of Madagascar.

Leonhardite (var. of laumontite), for Karl Cäsar von Leonhard (1779–1862), German mineralogist.

Leonite, for Leo Strippelmann, director of the salt works at Westerregeln, Germany.

Lepidocrocite, from Greek for *scale* and *fiber* or *thread,* in reference to the scaly or feathery habit.

Lepidolite, from Greek for *scale,* in allusion to the micaceous structure.

Lepidomelane (var. of biotite), from Greek words *scale* and *black,* in allusion to micaceous nature and color.

Lermontovite, for Mikhail Yurievich Lermontov (1814–1841), Russian writer and poet.

Letovicite, locality at Letovice, Moravia, Czechoslovakia.

Leucite, from Greek for *white,* in reference to its color.

Leucophanite, from Greek for *white* and *to appear,* because it shows a whitish reflection in certain lights.

Leucophoenicite, from Greek for *white* or *pale* and *purple-red,* in allusion to its color.

Leucophosphite, from Greek for *white* and from the element phosphorus, in allusion to its color and composition.

Leucosphenite, from Greek for *white* and *wedge,* in allusion to its color and wedge-shaped crystals.

Leucoxene (mixture of alteration products of ilmenite), from Greek for *white* and *stranger,* in allusion to its color and secondary nature.

Levyne, for Armand Lévy (1794–1841), French mathematician and natural scientist.

Lewisite, for William James Lewis (1847–1926), professor, Cambridge, England.

Lewistonite, locality near Lewiston, Utah.

Liberite, from composition, *li*thium and *be*ryllium.

Libethenite, locality at Libethen, near Neusohl, Rumania.

Liddicoatite, for Richard Thomas Liddicoat (1918–   ), American gemologist, Gemological Institute of America, Los Angeles.

Liebenbergite, for W. R. Liebenberg, Deputy Director-General of the National Institute for Metallurgy, South Africa.

Liebigite, for Justus von Liebig (1803–1873), German chemist.

Likasite, locality at the Likasi copper mine, Zaire.

Lillianite, for the Lillian Mining Company, Printerboy Hill, near Leadville, Colorado.

Lime, an Old English word for the substance; related to Dutch *lijm* and Latin *limus* (= *mud*) and *linere* (= *to smear*).

Limonite (general term for hydrous iron oxides, mostly goethite), from Greek for *meadow,* in allusion to its occurrence in bogs and swamps.

Linarite, from the supposed locality at Linares, Jaen province, Spain.

Lindackerite, for Joseph Lindacker, Austrian chemist, who made the first analysis.

Lindgrenite, for Waldemar Lindgren (1860–1939), Swedish-American mining geologist and teacher.

Lindströmite, for Gustaf Lindström (1838–1916), Swedish mineral analyst, of the Riksmuseet.

Linnaeite, for Carolus Linnaeus (1707–1778), Swedish botanist and taxonomist.

Liottite, for Luciano Liotti, Italian mineral collector, who supplied the original specimen from Tuscany.

Lipscombite, for William Nunn Lipscomb, Jr. (1919–   ), American physical chemist and crystallographer, a Nobel laureate, Harvard University.

Liroconite, from Greek for *pale* and *powder,* in allusion to its light blue to green streak.

Liskeardite, locality at Liskeard, Cornwall, England.

Litharge, from a Greek word given by Dioscorides to a material obtained in the process of separating lead from silver by fire metallurgy, the mineral being lead oxide.

Lithidionite, from Greek *lithidion* (= *pebble*), apparently in reference to the small lapilli fragments in which the mineral was found.

Lithiophilite, from lithium and the Greek word for *friend,* in allusion to composition.

Lithiophorite, from lithium and Greek for *to bear*, in allusion to the presence of the element in its composition.

Lithiophosphate, from composition, *lithi*um and *phosphate*.

Liveingite, for George D. Liveing (1827–1924), British chemist, Cambridge University.

Livingstonite, for David Livingstone (1813–1873), Scottish doctor and missionary, and African explorer.

Lizardite, locality at The Lizard in Cornwall, England.

Loellingite, locality at Lölling, near Hüttenberg, in Carinthia, Austria.

Loeweite, for Alexander Loewe (1808–1846), German chemist.

Lokkaite, for Lauri Lokka, Finnish chemist, Geological Survey of Finland.

Lomonosovite, for Mikhail V. Lomonosov (1711–1765), Russian naturalist and mineralogist.

Lonsdaleite, for Kathleen (Yardley) Lonsdale (1903–1971), British crystallographer.

Loparite, from Russian name for the Lapp inhabitants of the Kola Peninsula, Soviet Union.

Lopezite, for Emiliano Lopez, a mineral collector associated with nitrate mining, of Iquique, Chile.

Lorandite, for Eötvös Lorand, Hungarian physicist of Budapest.

Loranskite, for Apollonie Mikhailovich Loranski (1847–?), Inspector of the Mining Institute, St. Petersburg, Russia.

Lorenzenite, for Johannes Theodor Lorenzen (1855–1884), Danish mineralogist, student of minerals of Greenland.

Lorettoite, locality at Loretto, Tennessee.

Loseyite, for Samuel R. Losey (1833?–1906?), mineral collector, of Franklin, New Jersey.

Loughlinite, for Gerald Francis Loughlin (1880–1946), Chief Geologist of the U.S. Geological Survey, who did much to encourage research in mineralogy.

Lovdarite, from Russian and signifies "gift of Lovozero," in reference to locality on Karnasurt Mountain, Lovozero alkaline massif, Kola Peninsula, Soviet Union.

Loveringite, for John Francis Lovering (1930–   ), Australian geochemist, Australian National University, Canberra.

Lovozerite, locality at Lovozero alkaline massif, Kola Peninsula, Soviet Union.

Ludlamite, for Henry Ludlam (1824–1880), English mineralogist and collector of London.

Ludlockite, for Frederick *Lud*low Smith, III, and Charles *Lock*e Key,

American mineral collectors of New Jersey, who supplied the original specimens for study.

Ludwigite, for Ernst Ludwig (1842–1915), Austrian chemist, University of Vienna.

Lueneburgite, locality at Lüneburg, Hannover, Germany.

Lueshite, locality at Lueshe, north of Goma, Zaire.

Luetheite, for Ronald D. Luethe, American geologist of Douglas, Arizona, who first found the mineral.

Lusungite, locality near the Lusungu River, Zaire.

Luzonite, locality at Mancayan, Luzon, Philippine Islands.

M

Macdonaldite, for Gordon Andrew Macdonald (1911–1978), American volcanologist, University of Hawaii.

Macedonite, locality in province of Macedonia, Yugoslavia.

Machatschkiite, for Felix Karl Ludwig Machatschki (1895–1970), Austrian mineralogist, successively at Tübingen, Munich, and Vienna.

Mackayite, for John William Mackay (1831–1902), Irish-American mine operator on the Comstock Lode and benefactor of the Mackay School of Mines, University of Nevada.

Mackinawite, locality at Mackinaw mine, Snohomish County, Washington.

Madocite, locality in marble near Madoc, Ontario.

Magadiite, locality at Lake Magadi, Kenya.

Magbasite, from composition, *mag*nesium, *ba*rium, and *si*licate.

Maghemite, from first syllables of *mag*netite and *hem*atite, in allusion to the magnetism and composition.

Magnesioarfvedsonite, from composition, like arfvedsonite, but with predominant magnesium.

Magnesioaxinite, from composition, like axinite, but with predominant magnesium.

Magnesiochromite, from composition, like chromite, but with predominant magnesium.

Magnesiocopiapite, from composition, like copiapite, but with predominant magnesium.

Magnesioferrite, from composition, *magnesi*um and *ferr*ic iron.

Magnesiokatophorite, from composition, like katophorite, but with predominant magnesium.

Magnesioriebeckite, from composition, like riebeckite, but with predominant magnesium.

Magnesite, from composition, contains magnesium.

Magnesium astrophyllite, from composition, the magnesium content, and its close similarity to astrophyllite.

Magnesium-chlorophoenicite, from composition, like chlorophoenicite, but with predominant magnesium.

Magnesium-zippeite, from composition, like zippeite, but with predominant magnesium.

Magnetite, an ancient term generally attributed to the locality Magnesia, a district in Thessaly, Greece; some authorities derive it from Magnes, a shepherd, who first discovered the mineral on Mount Ida by noting that the nails of his shoes and the iron ferrule of his staff clung to a rock.

Magnetoplumbite, in allusion to its magnetic property and the presence of lead (*plumb*um) in its composition.

Magniotriplite, from composition, like triplite, but with predominant magnesium.

Magnocolumbite, from composition, like manganocolumbite or ferrocolumbite, but with predominant magnesium.

Magnophorite, from composition, magnesium, and Greek word for *to bear,* in allusion to the presence of the element.

Magnussonite, for Nils Harald Magnusson (1890–   ), Director of the Geological Survey of Sweden.

Majakite, locality at Majak mine, Talnakh deposit, Noril'sk region, Soviet Union.

Majorite, for Alan Major, who participated in studies on the synthesis of garnet from pyroxenes.

Makatite, from the Masai (Kenya) word, *emakat,* which means *soda,* in allusion to the high sodium content of the mineral.

Mäkinenite, for Eero Mäkinen, Finnish geologist and former president of the Outokumpu Company, Finland.

Malachite, from the Greek for *mallow,* in allusion to its green color.

Malacon (var. of zircon), from Greek for *soft.*

Malanite, for Chinese type locality, no specific name given.

Malayaite, locality at Perak, Malaya, Malaysia.

Maldonite, locality at Maldon, Victoria, Australia.

Malladrite, for Alessandro Malladra (1868–1945), Italian volcanologist, director of the Vesuvius Observatory.

Mallardite, for François Ernest Mallard (1833–1894), French mineralogist and crystallographer.

Manandonite, locality near the Manandona River, Madagascar.

Manasseite, for Ernesto Manasse (1875–1922), Italian chemist, mineralogist, and petrographer, University of Florence.

Mandarinoite, for Joseph Anthony Mandarino (1929–), American-Canadian mineralogist and curator, Royal Ontario Museum, Toronto.

Manganaxinite, from composition, like axinite, but with predominant manganese.

Manganbabingtonite, from composition, like babingtonite, but with predominant manganese.

Manganbelyankinite, from composition, like belyankinite, but with predominant manganese.

Manganberzeliite, from composition, like berzeliite, but with predominant manganese.

Manganese-hoernesite, from composition, like hoernesite, but with predominant manganese.

Manganese-shadlunite, from composition, like shadlunite, but with predominant manganese.

Manganhumite, from composition, like humite, but with predominant manganese.

Manganite, from composition, contains manganese.

Mangan-neptunite, from composition, like neptunite, but with predominant manganese.

Manganocolumbite, from composition, like magnocolumbite or ferrocolumbite, but with predominant manganese.

Manganolangbeinite, from composition, like langbeinite, but with predominant manganese.

Manganosite, from composition, contains manganese.

Manganostibite, from composition, manganese and antimony (*stib*ium).

Manganotantalite, from composition, like ferrotantalite, but with predominant manganese.

Manganpyrosmalite, from composition, like pyrosmalite, but with predominant manganese.

Manjiroite, for Manjiro Watanabe (1891–), professor, Tohoku University, Japan.

Mansfieldite, for George Rogers Mansfield (1875–1947), American geologist, U.S. Geological Survey.

Marcasite, a term derived from either an Arabic or Moorish word applied to pyrite and other substances.

Margarite, from Greek for *pearl,* in allusion to its aggregates of lamellae with a pearly luster.

Margarosanite, from Greek for *pearl* and *tablet,* in allusion to its luster and structure.

Marialite, for Maria Rosa, wife of Gerhard vom Rath (1830–1888), German mineralogist.

Maricite, for Luka Marić, Yugoslavian mineralogist, University of Zagreb.

Marokite, locality in Morocco (Maroc).

Marrite, for John Edward Marr (1857–1933), British geologist, Cambridge.

Marshite, for C. W. Marsh, who first described the mineral.

Marthozite, for Aimé Marthoz (1894–1962), Director-general of the Union Minière du Haut-Katanga, Africa.

Mascagnite, for Paolo Mascagni (1752–1815), Italian anatomist, University of Siena, who first described the natural salt.

Massicot, a French name for oxide of lead, related to Spanish *mazacote;* a long history with Arabic origins.

Masutomilite, for Kazunosuke Masutomi, Japanese amateur mineralogist and mineral collector.

Masuyite, for Gustave Masuy (?–1945), Belgian geologist.

Matildite, locality at Matilda mine, near Morococha, Peru.

Matlockite, locality at Matlock, Derbyshire, England.

Matraite, locality in the Matra Mountains, Hungary.

Mattagamite, locality near Mattagami Lake, Quebec.

Matteuccite, for Vittorio Matteucci (1862–1909), Italian volcanologist, director of the Vesuvius Observatory.

Matulaite, for Margaret Matula, amateur mineralogist of Allentown, Pennsylvania, who provided samples used for the study of the mineral.

Maucherite, for Wilhelm Maucher (1879–1930), German mineral dealer, Munich.

Maufite, for Herbert Brantwood Maufe (1879–1946), British geologist, director of the Southern Rhodesian Geological Survey.

Mawsonite, for Douglas Mawson (1882–1958), English-Australian geologist and Antarctic explorer.

Mayenite, locality near Mayen, Eifel district, Rhineland-Palatinate, West Germany.

Mazzite, for Fiorenzo Mazzi, Italian mineralogist, University of Pavia.

Mboziite, locality at the Mbozi syenite-gabbro complex in southwest Tanganyika.

Mcallisterite, for James Franklin McAllister (1911–), American geologist, U.S. Geological Survey, who first noticed the mineral in Death Valley, California.

Mcconnellite, for Richard Bradford McConnell, geologist, director of the British Guiana Geological Survey.

Mcgovernite, for J. J. McGovern (?–1915), American mineral collector of Franklin, New Jersey.

Mckelveyite, for Vincent Ellis McKelvey (1916–), American geologist, director of the U.S. Geological Survey.

Mckinstryite, for Hugh Exton McKinstry (1896–1961), American economic geologist, Harvard University.

Meerschaum (syn. for sepiolite), a German word meaning *sea-froth*, alluding to its lightness and color.

Meionite, from Greek for *less*, because the pyramid is less acute than in vesuvianite, with which it was associated.

Meixnerite, for Heinrich Herman Meixner (1908–), Austrian mineralogist, University of Salzburg.

Melanite (var. of andradite), from Greek for *black*, in allusion to its color.

Melanocerite, from Greek for *black* and the element cerium, in allusion to its color and composition.

Melanophlogite, from Greek for *black* and *to be burned*, because the mineral turns black superficially when heated with a blowpipe flame.

Melanostibite, from Greek for *black* and the element antimony (*stib*ium), in allusion to color and composition.

Melanotekite, from Greek for *black* and *to melt*, in allusion to the black bead it forms when fused in a blowpipe flame.

Melanovanadite, from Greek for *black* and the element vanadium, in allusion to its color and composition.

Melanterite, from Greek for a black metallic dye; used for copperas, the substance having formed as a byproduct in obtaining copper (although it is an iron sulfate).

Melilite, from Greek for *honey*, in reference to the common yellow color of the mineral.

Meliphanite, from Greek for *honey* and *to appear*, for its honey-yellow color.

Melkovite, for Vyacheslav Gavrilovich Melkov, Russian mineralogist.

Mellite, from Greek for *honey*, in reference to its honey-yellow color.

Melonite, locality at Melones mine, Calaveras County, California.

Melonjosephite, for Joseph Melon, Belgian mineralogist, University of Liège.

Mendipite, locality in Mendip Hills, Somersetshire, England.

Mendozite, locality at Mendoza, Argentina.

Meneghinite, for Giuseppi Meneghini (1811–1889), of Pisa, Italy, who first observed the species.

Mercallite, for Giuseppi Mercalli (1850–1914), Italian volcanologist, director of the Vesuvius Observatory.

Mercury, from Latin *Mercurius,* in mythology the herald or messenger of the gods, in allusion to the metal's mobility.

Merenskyite, for Hans Merensky (1871–1952), South African geologist, whose name was also given to the Merensky reef, western Bushveld, Transvaal, where the mineral occurs.

Merlinoite, for Stefano Merlino (1938–   ), Italian crystallographer, University of Pisa.

Merrihueite, for Craig M. Merrihue, Smithsonian Astrophysical Observatory.

Mertieite, for John Beaver Mertie, Jr. (1888–   ), American geologist, U.S. Geological Survey, who studied the platinum placers of Goodnews Bay, Alaska.

Merwinite, for Herbert Eugene Merwin (1878–1963), American geologist and mineralogist, Carnegie Institution, Washington, D.C.

Mesolite, from Greek for *middle,* in allusion to its intermediate chemistry between natrolite and scolecite.

Messelite, locality at Messel, Hesse, Germany.

Meta-aluminite, from Greek *meta,* meaning *along with,* and aluminite; indicates a lower hydrate than aluminite.

Meta-alunogen, from Greek *meta* and alunogen; indicates a lower hydrate than alunogen.

Meta-ankoleite, from Greek *meta,* indicating a low hydration state, and locality in Ankole district, Uganda.

Meta-autunite, from Greek *meta* and autunite; indicates a lower hydrate than autunite.

Metaborite, from composition, the same as metaboric acid.

Metacalciouranoite, from Greek *meta* and calciouranoite; indicates a lower hydrate than calciouranoite.

Metacinnabar, from Greek *meta* and cinnabar, because it was found associated with cinnabar.

Metadelrioite, from Greek *meta* and delrioite; indicates a lower hydrate than delrioite.

Metaheinrichite, from Greek *meta* and heinrichite; indicates a lower hydrate than heinrichite.

Metahewettite, from Greek *meta* and hewettite; in this case the two minerals are dimorphous.

Metahohmannite, from Greek *meta* and hohmannite; indicates a lower hydrate than hohmannite.

Metajennite, from Greek *meta* and jennite; indicates a lower hydrate than jennite.

Metakahlerite, from Greek *meta* and kahlerite; indicates a lower hydrate than kahlerite.

Metakirchheimerite, from Greek *meta* (indicating a low hydration state) and for Franz Waldemar Kirchheimer (1911–   ), German geologist, director of the Geologisches Landesamt for Württemberg-Baden.

Metalodevite, from Greek *meta* (indicating a low hydration state) and locality at Lodève, France.

Metanovacekite, from Greek *meta* and novacekite; indicates a lower hydrate than novacekite.

Metarossite, from Greek *meta* and rossite; indicates a lower hydrate than rossite.

Metaschoderite, from Greek *meta* and schoderite; indicates a lower hydrate than schoderite.

Metaschoepite, from Greek *meta* and schoepite; indicates a lower hydrate than schoepite.

Metasideronatrite, from Greek *meta* and sideronatrite; they differ only in total water content.

Metastibnite, from Greek *meta* and stibnite; they have the same composition, but metastibnite is amorphous.

Metatorbernite, from Greek *meta* and torbernite; indicates a lower hydrate than torbernite.

Metatyuyamunite, from Greek *meta* and tyuyamunite; indicates a lower hydrate than tyuyamunite.

Meta-uranocircite, from Greek *meta* and uranocircite; indicates a lower hydrate than uranocircite.

Meta-uranopilite, from Greek *meta* and uranopilite; indicates a lower hydrate than uranopilite.

Meta-uranospinite, from Greek *meta* and uranospinite; indicates a lower hydrate than uranospinite.

Metavandendriesscheite, from Greek *meta* and vandendriesscheite; indicates a lower hydrate than vandendriesscheite.

Metavanuralite, from Greek *meta* and vanuralite; indicates a lower hydrate than vanuralite.

Metavariscite, from Greek *meta* and variscite; in this case the two minerals are dimorphous.

Metavauxite, from Greek *meta* and vauxite; in this case it is a higher hydrate than vauxite.

Metavivianite, from Greek *meta* and vivianite; in this case the two minerals are dimorphous.

Metavoltine, from Greek *meta* (here meaning *with*) and voltine, because it was found associated with voltine in original locality.

Metazellerite, from Greek *meta* and zellerite; indicates a lower hydrate than zellerite.

Metazeunerite, from Greek *meta* and zeunerite; indicates a lower hydrate than zeunerite.

Meyerhofferite, for Wilhelm Meyerhoffer (1864–1906), German chemist, who synthesized the mineral.

Meymacite, locality at Meymac, Corrèze, France.

Miargyrite, from Greek words for *less* and *silver,* in allusion to the fact that it contains less silver than the ruby silver minerals.

Mica (mineral group name), apparently from Latin *micare,* meaning to shine; or perhaps from Latin *mica* (= *a crumb* or *grain*), if applied to scales.

Michenerite, for Charles Edward Michener (1907–), Canadian geologist, Canadian Nickel Company, Toronto, who described the mineral.

Microcline, from Greek for *little* and *slanted,* in reference to the slight variation of the cleavage angle from 90°.

Microlite, from the Greek word for *little* or *small,* in allusion to the minute size of the crystals from the original locality.

Microsommite, from the Greek word for *small,* in allusion to the minute prismatic crystals, and locality at Monte Somma, Vesuvius, Italy, where it was discovered.

Miersite, for Henry Alexander Miers (1858–1942), English mineralogist, Oxford University.

Milarite, named for Val Milar, Switzerland, in spite of the fact that this was not the correct locality for the original specimens (they came from Val Giuf).

Millerite, for William Hallowes Miller (1801–1880), British mineralogist, Cambridge University.

Millisite, for F. T. Millis, of Lehi, Utah, who found the first specimens.

Millosevichite, for Federico Millosevich (1875–1942), Italian mineralogist, University of Rome.

Mimetite, from Greek word meaning *imitator,* in allusion to its resemblance to pyromorphite.

Minasragrite, locality at Minasragra, near Cerro de Pasco, Peru.

Minguzzite, for Carlo Minguzzi, Italian mineralogist, University of Pavia.

Minium, from Latin *minium,* derived from an Iberian word for cinnabar, the Romans getting their cinnabar from Spain; although originally applied to cinnabar, the name is now used for a red lead oxide.

Minnesotaite, locality in the state of Minnesota.

Minyulite, locality near Minyulo Well, Dandaragan, Western Australia.

Mirabilite, from Latin *sal mirabile* (= *wonderful salt*); named by J. R. Glauber (1603–1668) who was surprised at its formation during an experiment.

Misenite, locality on Cape Miseno near Naples, Italy.

Miserite, for Hugh Dinsmore Miser (1884–1969), American geologist, U.S. Geological Survey.

Mitridatite, locality at Mitridat mountain, Kerch Peninsula, Crimea, Soviet Union.

Mitscherlichite, for Eilhard Mitscherlich (1794–1863), German crystallographer and chemist, who first prepared the synthetic compound.

Mixite, for A. Mixa, a mining official at Joachimsthal, Bohemia, where mineral occurs.

Mizzonite, from Greek word meaning *greater*, in allusion to its larger *c/a* ratio in comparison with meionite, a related mineral.

Moctezumite, locality at Moctezuma mine, near Moctezuma, east-central Sonora, Mexico.

Mohrite, for Karl Friedrich Mohr (1806–1879), German analytical chemist; the synthetic compound was long known as Mohr's salt.

Moissanite, for Ferdinand F. Henri Moissan (1852–1907), French chemist, who discovered the natural occurrence.

Moluranite, from composition, *mol*ybdenum and *uran*ium.

Molybdenite, from composition, *molybden*um, a word derived from Greek *molybdos* (= *lead*).

Molybdite, from composition, *molybd*ic oxide.

Molybdomenite, from Greek for *lead* and *moon*, in allusion to its composition which contains lead and selenium (element name is derived from a different Greek word meaning *moon*).

Molybdophyllite, from Greek words for *lead* and *leaf*, in allusion to its composition and foliated structure.

Molysite, from Greek word for *a stain*, in allusion to its staining of lavas on which it occurs.

Monazite, from Greek word meaning *to be solitary*, in allusion to the rarity of the mineral.

Moncheite, locality at Monchegorsk deposit and Monche Tundra, Soviet Union.

Monetite, locality on Island of Moneta, in the Caribbean Sea.

Monimolite, from Greek word for *stable*, because it is decomposed chemically with great difficulty.

Monohydrocalcite, from composition, *calcite* with one (*mono*) water (*hydro*).

Monsmedite, from the Latin name *Mons Medius* for Baia Sprie, Baia Mare region, Rumania, the locality of the mineral.

Montanite, locality in the state of Montana.

Montbrayite, locality in Montbray township, Abitibi County, Quebec.

Montebrasite, locality at Montebras, Creuse, France.

Monteponite, locality near Monte Ponti, Sardinia.

Montgomeryite, for Arthur Montgomery (1909–   ), American geologist and mineralogist, Lafayette College, Easton, Pennsylvania, who first recognized the mineral as a new species.

Monticellite, for Teodoro Monticelli (1759–1845), Italian mineralogist.

Montmorillonite, locality at Montmorillon, Vienne, France.

Montroseite, locality in Montrose County, Colorado.

Montroydite, for Montroyd Sharp, one of the owners of the mine in Brewster County, Texas, where the mineral occurs.

Mooihoekite, locality on Mooihoek Farm, Lydenburg District, Transvaal, South Africa.

Mooreite, for Gideon E. Moore (1842–1895), American chemist, early student of minerals from Franklin, New Jersey.

Moorhouseite, for Walter Wilson Moorhouse (1913–1969), Canadian geologist, University of Toronto.

Moraesite, for Luciano Jacques de Moraes (1896–   ), Brazilian geologist and mineralogist.

Mordenite, locality near Morden, King's County, Nova Scotia.

Morenosite, for a Señor Moreno (19th century), of Spain.

Morganite (var. of beryl), for John Pierpont Morgan (1837–1913), American banker, philanthropist, and collector of gems and art.

Morinite, for a Mr. Morineau, director of the tin mine at Montebras, Plateau Central, France.

Morion (var. of quartz), from Latin *mormorion,* an old name used for the mineral by Pliny.

Mosandrite, for Carl Gustav Mosander (1797–1858), Swedish chemist and mineralogist, who researched and discovered several rare earth elements.

Moschellandsbergite, locality at Moschellandsberg (Landsberg, near Ober-Moschel), Bavaria.

Mosesite, for Alfred J. Moses (1859–1920), American mineralogist, Columbia University, New York.

Mossite, locality near Moss, Norway.

Mottramite, locality at Mottram, St. Andrew, Cheshire, England.

Motukoreaite, locality at Brown's Island (Motukorea), within Waitemata Harbor, Auckland, New Zealand.

Mounanaite, locality at Mounana, Haut-Ogoue, Gabon, Africa.

Mountainite, for Edgar Donald Mountain, geologist at Rhodes University, Grahamstown, Union of South Africa.

Mourite, from composition, *mo*lybdenum and *ur*anium.

Mpororoite, locality at Mpororo tungsten deposit, Kigezi district, Uganda.

Mroseite, for Mary E. Mrose (1910–), American mineralogist, U.S. Geological Survey.

Muirite, for John Muir (1838–1914), American geologist and explorer, who made early observations in geology in the Sierra Nevada Mountains, California.

Mukhinite, for A. S. Mukhin, geologist of the Western Siberia Geological Administration, Soviet Union.

Mullite, locality on the Island of Mull, Scotland.

Murataite, for Kiguma Jack Murata (1909–), geochemist of the U.S. Geological Survey, Menlo Park, California.

Murdochite, for Joseph Murdoch (1890–1973), American mineralogist, University of California at Los Angeles.

Murmanite, apparently for locality near Murmansk, Kola. Peninsula, Soviet Union.

Muscovite, from *Muscovy glass*, name used when mineral was first described from the Russian province of Muscovy.

Muskoxite, locality at Muskox Intrusion, Northwest Territories, Canada.

Muthmannite, for Friedrich Wilhelm Muthmann (1861–1913), German chemist and crystallographer, of Munich.

N

Nacrite, from French *nacre, mother-of-pearl*, in allusion to its appearance.

Nadorite, locality at Djebel Nador, Constantine, Algeria.

Nagelschmidtite, for Gunther Nagelschmidt, chemist who lived in England.

Nagyagite, locality at Nagyág in Transylvania, Rumania.

Nahcolite, from composition, *Na, H, C, O*.

Nakaseite, locality at the Nakase mine, Japan.

Nakauriite, locality at Nakauri, Achi Prefecture, Japan.

Nambulite, for Matsuo Nambu, professor, Tohoku University, Japan.

Nanlingite, locality in Nan Ling, in southern China.

Nantokite, locality at Nantoko, Chile.

Narsarsukite, locality in pegmatite at Narsarsuk, southern Greenland.

Nasinite, for Raffaello Nasini (1854–1931), Italian chemist.

Nasledovite, for Boris Nikolaevich Nasledov, Russian geologist, investigator of mineral resources of Kara-Mazar.

Nasonite, for Frank Lewis Nason (1856–1928), American geologist, Geological Survey of the State of New Jersey.

Natisite, from composition, *Na, Ti, Si.*

Natroalunite, from composition, like alunite, but with predominant sodium (*natr*ium).

Natrochalcite, from composition, sodium (*natr*ium) and copper (*chalk*os).

Natrofairchildite, from composition, like fairchildite, but with predominant sodium (*natr*ium).

Natrojarosite, from composition, like jarosite, but with predominant sodium (*natr*ium).

Natrolite, from Latin *natrium, sodium,* in allusion to the composition.

Natromontebrasite, from composition, like montebrasite, but with predominant sodium (*natr*ium).

Natron, ancient origin, from Greek *nitron* and Latin *natrium,* used to describe the mineral.

Natroniobite, from composition, sodium (*natr*ium) and *niob*ium.

Natrophilite, from composition, sodium (*natr*ium), and Greek for *friend,* in allusion to the presence of the element.

Natrophosphate, from composition, sodium (*natr*ium) and *phosphate.*

Natrosilite, from composition, sodium (*natr*ium) and *sil*icon.

Naujakasite, locality at Naujakasik, Tunugdliarfik fjord, southwestern Greenland.

Naumannite, for Karl Friedrich Naumann (1797–1873), German crystallographer and mineralogist.

Navajoite, for the Navajo Indians; found on the Navajo Indian Reservation, Apache County, Arizona.

Neighborite, for Frank Neighbor, American geologist, Sun Oil Company, Casper, Wyoming.

Nekoite, derived from *oken*ite by reversing the first four letters to become *neko;* the minerals are very similar.

Nenadkevichite, for Konstantin A. Nenadkevich (1880–1963), Russian mineralogist and geochemist.

Neotocite, from Greek for *new born,* alluding to the fact it is an alteration mineral.

Nepheline, from Greek for *cloud,* because when the mineral is immersed in acid it becomes cloudy.

Nephrite (var. of actinolite; a jade), from Latin *lapis nephriticus* (= *kidney stone*), because it was worn as a remedy for diseases of the kidney.

Nepouite, locality at Nepoui, New Caledonia.

Neptunite, from Neptune, in Roman mythology the god of the sea; so called because it was found with aegirine which was named for Ægir, the Scandinavian god of the sea.

Nesquehonite, locality at Nesquehoning, near Lansford, Carbon County, Pennsylvania.

Newberyite, for James Cosmo Newbery (1843–1895), Australian geologist of Melbourne.

Neyite, for Charles Stuart Ney (1918–1975), Canadian geologist, North Vancouver, British Columbia, in charge of early exploration of the deposit where the mineral was discovered.

Niccolite (syn. for nickeline), from *niccolum,* a Latinized form of the original German name *Kupfernickel* which meant "old Nick's copper." The mineral was believed to contain copper, but would not yield it; it is a nickel arsenide.

Nickel, from German *Nickel,* derived from the mineral name *Kupfernickel,* meaning *Old Nick's* or *the devil's copper;* the mineral, a nickel arsenide, was believed to contain copper, but would not yield it.

Nickelbloedite, from composition, like bloedite, but with predominant nickel.

Nickel hexahydrite, from composition, like hexahydrite, but with predominant nickel.

Nickeline, from German *Kupfernickel* and *Nickel;* see discussions under *niccolite* and *nickel.*

Nickel-iron, named for the two metals in the alloy.

Nickel-skutterudite, from composition, like skutterudite, but with predominant nickel.

Nickel-zippeite, from composition, like zippeite, but with predominant nickel.

Nifontovite, for Roman Vladimirovich Nifontov, Russian geologist.

Nigerite, locality in the Kabba Province, central Nigeria.

Niggliite, for Paul Niggli (1888–1953), Swiss mineralogist, University of Zürich.

Nimite, from the abbreviation of the National Institute of Metallurgy (NIM), South Africa.

Ningyoite, locality at Ningyo-toge mine, Tottori Prefecture, Japan.

Niningerite, for Harvey Harlow Nininger (1887–   ), American meteoriticist of Sedonia, Arizona.

Niobo-aeschynite, from composition, like aeschynite, but with predominant niobium.

Niobophyllite, from composition, niobium, and Greek word for *leaf,* in allusion to its micaceous nature.

Niocalite, from composition, *nio*bium and *calc*ium.

Nisbite, from composition, *Ni* and *Sb.*

Nissonite, for William H. Nisson (1912–1965), American amateur mineralogist of Petaluma, California, who made preliminary chemical tests on the mineral.

Niter, ancient origin, from Latin *nitrum,* from Greek *nitron,* from Hebrew *nether;* perhaps originally from Nitria, the name of a city in Upper Egypt, near which the mineral is found.

Nitrobarite, from composition, a *bar*ium *nitr*ate.

Nitrocalcite, from composition, a *calc*ium *nitr*ate.

Nitromagnesite, from composition, a *magnes*ium *nitr*ate.

Nobleite, for Levi Fatzinger Noble (1882–1965), American geologist, U.S. Geological Survey.

Nolanite, for Thomas Brennan Nolan (1901–   ), American geologist, Director of the U.S. Geological Survey.

Nontronite, locality in Arrondissement of Nontron, near the village of Saint Pardoux, France.

Norbergite, locality at Norberg, Sweden.

Nordenskiöldine, for Nils Adolf Erik Nordenskiöld (1832–1901), Swedish mineralogist and explorer.

Nordite, from Russian word for *north,* because of its northern origin in the Lovozero tundras, Soviet Union.

Nordstrandite, for Robert A. Van Nordstrand, Sinclair Research Laboratories, of Harvey, Illinois.

Norsethite, for Keith Norseth, engineering geologist of the trona mine at Westvaco, Sweetwater County, Wyoming.

Northupite, for Charles H. Northup (1861–?) American grocer, of San Jose, California, who found the first specimens.

Nosean, for Karl Wilhelm Nose (1753?–1835), German mineralogist, Brunswick.

Novacekite, for Radim Nováček (1905–1942), Czech mineralogist.

Novakite, for Jirí Novák (1902–   ), Czech mineralogist, Charles University, Prague.

Nowackiite, for Werner Nowacki (1909–   ), Swiss crystallographer, University of Berne.

Nsutite, locality at Nsuta, Ghana.

Nuffieldite, for Edward Wilfrid Nuffield (1914–   ), Canadian mineralogist, University of Toronto.

Nyerereite, for Julius K. Nyerere (1922–   ), former teacher and President of Tanzania, the country where the mineral was discovered.

O

Obruchevite, for Vladimir Afanasevich Obruchev (1863–1956), Russian geologist.

Offretite, for Albert Jules Joseph Offret (1857–?), professor, Lyons, France.

Okenite, for Lorenz Oken (1779–1851), German naturalist, Munich.

Oldhamite, for Thomas Oldham (1816–1878), director of the Indian Geological Survey (1850–1876).

Oligoclase, from Greek words meaning *little* and *fracture,* since it was believed to have less perfect cleavage than the related albite.

Olivenite, named in allusion to its olive-green color.

Olivine, named in allusion to its olive-green color.

Olmsteadite, for Milo Olmstead, American amateur mineral micromounter, of Rapid City, South Dakota, who first noticed the mineral.

Olsacherite, for Juan A. Olsacher (1903–1964), Argentine mineralogist, Córdoba University.

Olshanskyite, for Yakov Iosifovich Olshanskii (1912–1958), Soviet physical geochemist.

Omphacite, from the Greek for *unripe grape,* in allusion to its characteristic green color.

Onoratoite, for Ettore Onorato (1899–   ), Italian mineralogist.

Onyx (var. of quartz or calcite), from an ancient Greek word meaning claw, fingernail, hoof, veined gem.

Oosterboschite, for M. R. Oosterbosch, active in the development of the mines in Katanga where the mineral was discovered.

Opal, from Sanskrit *upala* (= *stone* or *precious stone*).

Orcelite, for Jean Orcel (1896–   ), French mineralogist.

Ordonezite, for Ezequiel Ordóñez (1867–1950), Mexican geologist, head of the Instituto de Geologia de Mexico.

Oregonite, locality in the state of Oregon.

Orientite, locality in Oriente, Cuba.

Orpheite, from Greek *Orpheus,* the mythological musician in the Rhodope Mountains, Bulgaria, the locality of the mineral.

Orpiment, from Latin *auripigmentum* (= *golden paint*), in allusion to its color and early use.

Orthoantigorite, from crystal system, the *ortho*rhombic dimorph of antigorite.

Orthochamosite, from crystal system, the *ortho*rhombic dimorph of chamosite.

Orthochrysotile, from crystal system, the *ortho*rhombic dimorph of chrysotile.

Orthoclase, from Greek for *straight* and *fracture,* in reference to the cleavage angle, which is 90°.

Orthoericssonite, from crystal system, the *ortho*rhombic dimorph of ericssonite.

Orthoferrosilite, from crystal system, the *ortho*rhombic dimorph of clinoferrosilite.

Orthopinakiolite, from crystal system, the *ortho*rhombic dimorph of pinakiolite.

Osarizawaite, locality at Osarizawa copper mine, Akita Prefecture, Japan.

Osarsite, from composition, contains *os*mium and *ars*enic.

Osbornite, for George Osborne, who sent to London the specimen in which it was discovered.

Osmiridium, from composition, *osm*ium and *iridium.*

Osmium, from Greek word *osme, a smell.*

Osumilite, locality in Osumi, the name of an old province in Sakkabira, Kyusyu, Japan.

Otavite, locality at Tsumeb, near Otavi, South West Africa.

Ottemannite, for Joachim Ottemann (1914–   ), German mineralogist, Heidelberg.

Ottrelite (var. of chloritoid), locality in schist near Ottrez, a village in Belgium on the border of Luxemburg.

Otwayite, for Charles Otway, Gosnells, Western Australia, who was helpful in providing access to, and samples from, his mineral lease.

Ourayite, locality at Old Laut's mine, Ouray, Colorado.

Overite, for Edwin J. Over, Jr. (1905–1963), American mineral collector of Colorado Springs, Colorado, who collected specimens from the type locality.

Owyheeite, locality at Owyhee County, Idaho.

Oxammite, from composition, including *oxa*late and *amm*onium.

P

Pabstite, for Adolf Pabst (1899–   ), American mineralogist, University of California, Berkeley.

Pachnolite, from Greek word for *frost,* in allusion to its appearance.

Painite, for A. C. D. Pain, gem collector who first recognized the unusual nature of the original specimen.

Palermoite, locality at Palermo pegmatite, North Groton, New Hampshire.

Palladium, for the asteroid Pallas (a name of the Greek goddess Athena), discovered about the same time.

Palladoarsenide, from composition, a *pallad*ium *arsenide.*

Palladobismutharsenide, from composition, *pallad*ium, *bismuth,* and *ar-senic.*

Palladseite, from composition, *pallad*ium and *Se.*

Palmierite, for Luigi Palmieri (1807–1896), Italian volcanologist, director of the observatory on Vesuvius.

Palygorskite, locality "in der Paligorischen Distanz" of the second mine on the Popovka River, Urals, Soviet Union.

Pandaite, locality at Panda Hill, Tanganyika.

Panethite, for Friedrich Adolf Paneth (1887–1958), Director of the Max Planck Institute, Mainz, Germany, who contributed to the study of meteorites.

Paolovite, from composition, *pa*lladium and tin (Russian *olov*o).

Papagoite, for the Papago Indian tribe that once inhabited the region in which the mining center of Ajo, Arizona, is situated.

Para-alumohydrocalcite, from Greek *para* (= *beside*) and alumohydrocalcite; the minerals are closely related chemically, but this one contains more water.

Parabutlerite, from Greek *para* and butlerite; a dimorphous pair.

Paracelsian, from Greek *para* and celsian; a dimorphous pair.

Parachrysotile from Greek *para* and chrysotile; one of several polymorphic forms related to chrysotile.

Paracoquimbite, from Greek *para* and coquimbite; the minerals are polytypic.

Paracostibite, from Greek *para* and costibite; a dimorphous pair.

Paradamite, from Greek *para* and adamite; a dimorphous pair.

Paradocrasite, from Greek words for *contrary to expectation* (or *paradox*) and *mixture;* discovered during a study of dyscrasite, a name from Greek meaning *a bad mixture.*

Paragonite, from Greek for *to mislead*, because it was originally mistaken for talc.

Paraguanajuatite, from Greek *para* and guanajuatite; a dimorphous pair.

Parahilgardite, from Greek *para* and hilgardite; a dimorphous pair.

Parahopeite, from Greek *para* and hopeite; a dimorphous pair.

Parajamesonite, from Greek *para* and jamesonite; a dimorphous pair.

Parakeldyshite, from Greek *para* and keldyshite; the minerals were associated in nature.

Paralaurionite, from Greek *para* and laurionite; a dimorphous pair.

Paramelaconite, from Greek *para* (here meaning *near*) and melaconite (old name for tenorite) which is from Greek meaning *black* and *dust*.

Paramontroseite, from Greek *para* and montroseite; it forms from the oxidation of montroseite, and may be pseudomorphic after it.

Parapierrotite, from Greek *para* and pierrotite; the minerals are apparently dimorphous.

Pararammelsbergite, from Greek *para* and rammelsbergite; a dimorphous pair.

Paraschachnerite, from Greek *para* and schachnerite; the minerals are closely related and apparently dimorphous.

Paraschoepite, from Greek *para* and schoepite; the minerals are closely related chemically.

Paraspurrite, from Greek *para* and spurrite; a dimorphous pair.

Parasymplesite, from Greek *para* and symplesite; a dimorphous pair.

Paratacamite, from Greek *para* and atacamite; along with botallackite they are trimorphous.

Paratellurite, from Greek *para* and tellurite; a dimorphous pair.

Paravauxite, from Greek *para* and vauxite; in allusion to its chemical relationship to vauxite.

Paraveatchite, from Greek *para* and veatchite; the minerals are closely related but have different crystal lattices.

Pargasite, locality at Pargas, Finland.

Parisite, for J. J. Paris, proprietor of the mine at Muzo, north of Bogota, Columbia, where the mineral was discovered.

Parkerite, for Robert Lüling Parker (1893–1973), professor of mineralogy and keeper of the mineral collection of the Swiss Federal Institute of Technology, Zürich.

Parnauite, for John L. Parnau, American mineral collector of Sunnyvale, California, who contributed to the knowledge of the type locality in Nevada.

Parsettensite, locality at Parsettens Alp, Val d'Err, Grisons, Switzerland.

Parsonsite, for Arthur Leonard Parsons (1873–1957), Canadian mineralogist, University of Toronto.

Partzite, for A. F. W. Partz, who first recognized the mineral as a silver ore.

Parwelite, for Alexander Parwel, Swedish chemist, Swedish Natural History Museum, who analyzed the mineral.

Pascoite, locality in Pasco province, Peru.

Patronite, for Antenor Rizo-Patrón, Peruvian engineer, discoverer of the ore in which it was found.

Paulingite, for Linus Carol Pauling (1901–), American chemist and physicist, a Nobel laureate, Linus Pauling Institute of Science and Medicine, Menlo Park, California.

Paulmooreite, for Paul Brian Moore (1940–), American mineralogist, University of Chicago.

Pavonite, from Latin *pavo,* meaning *peacock,* for Martin Alfred Peacock (1898–1950), Canadian mineralogist.

Paxite, from Latin *pax,* meaning peace.

Pearceite, for Richard Pearce (1837–1927), American chemist and metallurgist, Denver, Colorado.

Pecoraite, for William Thomas Pecora (1913–1972), American geologist, director of the U.S. Geological Survey.

Pectolite, from Greek word meaning *compact,* in allusion to its structure.

Pekoite, locality at Peko mine, Tennant Creek gold field, Northern Territory, Australia.

Pellyite, locality near the headwaters of the Pelly River, Yukon Territory, Canada.

Penfieldite, for Samuel Lewis Penfield (1856–1905), American mineralogist and mineral chemist, Yale University.

Penikisite, for Gunar Penikis (1936–), a discoverer of the phosphate occurrence in the northeastern corner of the Yukon Territory, Canada.

Penkvilksite, from Lapp words *penk* (= *curly*), and *vilkis* (= *white*), in allusion to its appearance.

Pennantite, for Thomas Pennant (1726–1798), Welsh traveller, zoologist, and mineralogist.

Penninite (var. of clinochlore), locality in the Pennine Alps, Switzerland.

Penroseite, for Richard Alexander Fullerton Penrose, Jr. (1863–1931), American geologist, benefactor of the Geological Society of America.

Pentagonite, from Greek words meaning *five* and *angle;* it occurs in prismatic crystals twinned to form fivelings with star-shaped cross sections.

Pentahydrite, from composition, it is a pentahydrate (contains five waters).

Pentahydroborite, from composition, contains five (*penta*) waters (*hydro*) and is a *bor*ate.

Pentlandite, for Joseph Barclay Pentland (1797–1873), Irish natural scientist and traveller; involved in South American studies.

Percylite, for John Percy (1817–1889), English metallurgist.

Perhamite, for Frank C. Perham, American geologist and pegmatite miner, West Paris, Maine, who is dedicated to the recovery of fine mineral specimens.

Periclase, from Greek for *around* and *fracture,* alluding to the perfect cubic cleavage.

Peridot (var. of forsterite), from French *péridot,* of unknown origin.

Peristerite (var. of albite), from Greek for *pigeon,* the colors somewhat resembling those of the neck of a pigeon.

Perite, for Per Adolf Geijer (1886–?), Swedish geologist.

Perloffite, for Louis Perloff, American amateur mineralogist, of Tryon, North Carolina.

Permingeatite, for François Permingeat (1917–   ), French mineralogist, University of Toulouse.

Perovskite, for Count Lev Aleksevich von Perovski (1792–1856), Russian mineralogist, of St. Petersburg.

Perrierite, for Carlo Perrier (1886–1948), Italian mineralogist.

Perryite, for Stuart Hoffman Perry (1874–1957), American meteoriticist.

Petalite, from Greek for *leaf,* in allusion to its leaflike cleavage.

Petrovicite, locality in Petrovice selenium deposits of western Moravia, Czechoslovakia.

Petzite, for W. Petz, who first analyzed the mineral.

Pharmacolite, from Greek for *poison,* in allusion to its arsenic content.

Pharmacosiderite, from Greek for *poison,* in allusion to arsenic, and for *iron.*

Phenakite, from Greek for *deceiver,* in allusion to its having been mistaken for quartz.

Phengite (var. of muscovite), from Greek and Latin *phengites,* the name for transparent or translucent stones used by the ancients for windows.

Phillipsite, for William Phillips (1775–1829), British mineralogist, a founder of the Geological Society of London.

Phlogopite, from Greek for *firelike,* in allusion to reddish tinge which some specimens display.

Phoenicochroite, from Greek for *deep red* and *color,* in allusion to its color.

Phosgenite, from *phosgene,* the name for carbonyl chloride, $COCl_2$, because the mineral contains carbon, oxygen, and chlorine in addition to other elements.

Phosinaite, from composition, *phos*phate, *s*ilicon, and sodium (*na*trium).

Phosphammite, from composition, *phosph*ate and *amm*onium.

Phosphoferrite, from composition, *phosph*ate of iron (*ferr*um).

Phosphophyllite, from composition, *phosph*ate, and Greek word for *leaf,* in allusion to its perfect cleavage.

Phosphorrösslerite, from composition, like rösslerite, but with predominant phosphate.

Phosphosiderite, from composition, *phosph*ate of iron (*sider*os).

Phosphuranylite, from composition, *phosph*ate with *uranyl* group.

Pickeringite, for John Pickering (1777–1846), American linguist and philologist, of Massachusetts.

Picotite (var. of spinel), for Picot de la Peyrous (1744–1818), Inspector of Mines and natural historian at Toulouse, France, who described the rock in which the mineral was discovered.

Picotpaulite, for Paul Picot, Bureau de recherches géologiques et minières, Orléans, France.

Picromerite, from Greek for *bitter* and *part,* in allusion to its content of magnesia (salts of which are usually bitter).

Picropharmacolite, from Greek for *bitter* (in allusion to its magnesia content), and its chemical similarity to pharmacolite.

Piemontite, locality in Piemonte, Italy.

Pierrotite, for Roland Pierrot, Bureau de recherches géologiques et minières, Orléans, France.

Pigeonite, locality at Pigeon Point, Minnesota.

Pimelite, from Greek for *fatness,* from its greasy appearance and feel.

Pinakiolite, from Greek for *small tablet,* in allusion to the thin tabular habit.

Pinchite, for William W. Pinch, American mineral collector, of Rochester, New York, who recognized the mineral as a new species.

Pinnoite, for a Mr. Pinno, Chief Councillor of Mines, of Halle, Germany.

Pintadoite, locality at Cañon Pintado, San Juan County, Utah.

Pirssonite, for Louis Valentine Pirsson (1860–1919), American mineralogist and petrographer, Yale University.

Pitticite, from Greek for *pitch,* because it was earlier called pitch-ore, in allusion to its structure and luster.

Plagioclase (series of minerals in the feldspar group), from Greek for *oblique* and *fracture,* in reference to the oblique angle between its best cleavages.

Plagionite, from Greek for *oblique,* in allusion to the obliquity of its monoclinic crystals.

Plancheite, for a Mr. Planche, who furnished the African material for study.

Planerite, for Dimitri Ivanovich Planer (1820–1882), director of the copper mines of Gumeshevsk, in the Urals, Soviet Union, where the mineral was discovered.

Plasma (var. of quartz), from Greek *plasma,* used for things *formed* or *molded,* probably applied because the mineral was used for intaglios.

Platarsite, from composition, *plat*inum and *ars*enic.

Platiniridium, from composition, alloy of *platin*um and *iridium.*

Platinum, from Spanish *plata,* meaning *silver.*

Plattnerite, for K. F. Plattner (1800–1858), professor of metallurgy and assaying, Freiberg, Germany.

Platynite, from Greek word for *to broaden,* in allusion to its platy structure.

Playfairite, for John Playfair (1748–1819), Scottish mathematician and geologist, University of Edinburgh.

Pleonaste (var. of spinel), from Greek for *excess,* because faces other than the octahedron occurred on crystals of the mineral.

Plombierite, locality at Plombières, France; deposited from thermal water on brick and mortar of an old Roman aqueduct.

Plumalsite, from composition, lead (*plumb*um), *Al,* and *Si.*

Plumboferrite, from composition, lead (*plumb*um) and iron (*ferr*um).

Plumbogummite, from composition, lead (*plumb*um), and Latin *gummi* (= *gum*), in allusion to its appearance.

Plumbojarosite, from composition, like jarosite, but with predominant lead (*plumb*um).

Plumbonacrite, from composition, lead (*plumb*um), and French *nacre* (= *mother-of-pearl*), in allusion to its luster.

Plumbopalladinite, from composition, lead (*plumb*um) and *palladi*um.

Plumbopyrochlore, from composition, like pyrochlore, but with predominant lead (*plumb*um).

Poitevinite, for Eugene Poitevin, Canadian mineralogist, Geological Survey of Canada.

Polarite, locality in the Polar Urals, Soviet Union.

Pollucite, from Pollux, in Classical mythology the twin brother of Castor, named for its association with the mineral castor (old name for petalite).

Polybasite, from Greek for *many* and *base,* in allusion to the many metallic bases in its composition.

Polycrase, from Greek for *many* and for *mixture,* in allusion to the many rare elements it contains.

Polydymite, from Greek for *many* and *twin,* because the mineral is often observed in twinned crystals.

Polyhalite, from Greek for *many* and *salt,* in allusion to the several component salts present.

Polylithionite, from Greek for *many* or *much,* and the element lithium, apparently in allusion to its high lithium content.

Polymignyte, from Greek for *many* and for *mix,* in allusion to its complex composition.

Portlandite, from Portland cement, with which the synthetic compound was known to be associated.

Posnjakite, for Eugene Waldemar Posnjak (1888–1949), Russian-American geochemist, of the Geophysical Laboratory, Washington, D.C.

Potarite, locality in the Potaro River region, British Guiana.

Potash alum, from composition, the potassium member of the alum group.

Poubaite, for Zdenek Pouba, Czech economic geologist, Charles University, Prague.

Poughite, for Frederick Harvey Pough (1906–), American mineralogist, American Museum of Natural History, New York, who first studied the mineral from Honduras.

Powellite, for John Wesley Powell (1834–1902), American explorer and geologist, director of the U.S. Geological Survey.

Prase (var. of quartz), from Greek for *leek-green,* in allusion to its color.

Prehnite, for Hendrik von Prehn (1733–1785), Dutch military figure of Cape Town, South Africa, and the Netherlands, who collected the specimens.

Preobrazhenskite, for Pavla Ivanovich Preobrazhensk (1874–1944), Soviet geologist, investigated the salt deposits of the Soviet Union.

Priceite, for Thomas Price (1837?–?), Welsh-American metallurgist of San Francisco, California, who first analyzed the mineral.

Priderite, for Rex Tregilgas Prider (1910–), Australian geologist, University of Western Australia.

Probertite, for Frank Holman Probert (1876–1940), Dean of the Mining College, University of California.

Prosopite, from Greek for *mask,* in allusion to the deceptive (pseudomorphous) character of the mineral.

Proudite, for J. S. Proud, a director of Peko-Wallsend Ltd., the company responsible for the development of the Tennant Creek gold field, Northern Territory, Australia, where mineral was discovered.

Proustite, for Joseph Louis Proust (1754–1826), French chemist.

Przhevalskite, for Nikolai Mikhailovich Przhevalsky (1839–1888), Russian explorer of Central Asia.

Pseudo-autunite, from Greek *pseudo* (= *false*), and autunite, a mineral it resembles in some respects.

Pseudoboleite, from Greek *pseudo* (= *false*), and boleite, a mineral it resembles in some respects.

Pseudobrookite, from Greek *pseudo* (= *false*), and brookite, originally thought to be similar to it.

Pseudocotunnite, from Greek *pseudo* (= *false*), and cotunnite, a mineral it resembles in some respects.

Pseudolaueite, from Greek *pseudo* (= *false*), and laueite, a mineral it resembles and with which it is probably dimorphous.

Pseudomalachite, from Greek *pseudo* (= *false*), and malachite, a mineral with a similar physical appearance.

Psilomelane (general term for massive, hard manganese oxides), from Greek for *smooth* and *black,* in allusion to its appearance.

Pucherite, locality at Pucher shaft of the Wolfgang mine at Schneeberg, Saxony.

Pumpellyite, for Raphael Pumpelly (1837–1923), a pioneer student of the copper deposits of the Keweenaw Peninsula, Michigan, the locality from which it was originally described.

Purpurite, from Latin *purpura,* in allusion to its purple color.

Pyralspite (subgroup of the garnet group), from the initial letters of *pyr*ope, *al*mandine, and *sp*essartine.

Pyrargyrite, from Greek for *fire* and *silver,* in allusion to its color and composition.

Pyribole (collective name for the two mineral groups), from the mineral groups *pyr*oxene and amph*ibole.*

Pyrite, from Greek for *fire,* in allusion to the fact that when struck with steel it gives off brilliant sparks.

Pyroaurite, from Greek for *fire* and Latin for *gold* (*aur*um), in allusion to the goldlike appearance after being heated.

Pyrobelonite, from Greek for *fire* and *needle,* in allusion to the red color and acicular habit.

Pyrochlore, from Greek for *fire* and *green,* because some specimens of the mineral turn green after ignition.

Pyrochroite, from Greek for *fire* and *color*, because of the change of color upon ignition.

Pyrolusite, from Greek for *fire* and *to wash*, because it can be used to free glass of the colors due to iron by its oxidizing effect.

Pyromorphite, from Greek for *fire* and *form*, in allusion to the crystalline state it assumes on cooling from a fusion.

Pyrope, from Greek for *fiery*, in allusion to its fire-red color.

Pyrophanite, from Greek for *fire* and *to appear*, in allusion to its red color.

Pyrophyllite, from Greek for *fire* and *a leaf*, because it exfoliates when heated.

Pyrosmalite, from Greek for *fire* and *odor*, in allusion to its odor when heated.

Pyrostilpnite, from the Greek words *fire* and *shining*, in allusion to its color and luster.

Pyroxene (mineral group name), from Greek for *fire* and *stranger;* it was incorrectly believed to be out of place in igneous rocks.

Pyroxenoid (mineral group name), from pyroxene, in allusion to similarities between the two groups of minerals.

Pyroxferroite, from *pyrox*ene (or *pyrox*enoid) and *ferro*us iron; named for its relationship to pyroxmangite.

Pyroxmangite, from *pyrox*ene and *mang*anese; it it not a true pyroxene as originally supposed, but a pyroxenoid.

Pyrrhotite, from Greek for *redness*, alluding to the liveliness of its color.

Q

Quartz, apparently from the Saxon word *querkluftertz*, meaning *cross-vein ore*, which could easily have become condensed to *querertz* and then to *quartz;* the idea is supported by the old Cornish name for crystalline silica which means *cross-course-spar;* in any case the word is from German and of ancient derivation.

Quenselite, for Percy Dudgeon Quensel (1881–1966), Swedish mineralogist and petrologist, of the University of Stockholm.

Quenstedtite, for Friedrich August Quenstedt (1809–1889), German mineralogist.

Quetzalcoatlite, from Quetzalcoatl, the Toltec and Aztec Indian god of the sea, in allusion to the blue color of the mineral.

R

Rabbittite, for John Charles Rabbitt (1907–1957), American geologist and geochemist, U.S. Geological Survey.

Raguinite, for Eugène Raguin (1900–), French geologist, Ecole Nationale Supérieure de Mines, Paris.

Raite, named in honor of the international collective of scientists on the voyage of the papyrus ship *Ra* (1969–1970), under the leadership of Thor Heyerdahl.

Ralstonite, for J. Grier Ralston, Norristown, Pennsylvania, who first observed the mineral.

Ramdohrite, for Paul Ramdohr (1890–), German mineralogist, Berlin.

Rameauite, for Jacques Rameau (?–1960), French prospector, who discovered the deposit where the mineral occurs.

Rammelsbergite, for Karl Friedrich Rammelsberg (1813–1899), German mineral chemist, Berlin.

Ramsdellite, for Lewis Stephen Ramsdell (1895–1975), American mineralogist, University of Michigan, Ann Arbor.

Rancieite, locality at Rancié, near Vicdessos, Ariège, France.

Rankamaite, for Kalervo Rankama (1913–), Finnish geochemist, University of Helsinki.

Rankinite, for George Atwater Rankin (1884–?), American physical chemist, Geophysical Laboratory, Washington, D.C., who first synthesized the compound.

Ransomite, for Frederick Leslie Ransome (1868–1935), American mining geologist, California Institute of Technology.

Raspite, for Charles Rasp (1846–1907), German-Australian prospector, the discoverer of the Broken Hill mines, New South Wales, Australia.

Rasvumite, locality in pegmatites of the Rasvumchorr apatite deposits, Khibina massif, Kola Peninsula, Soviet Union.

Rathite, for Gerhard vom Rath (1830–1888), German mineralogist of Bonn.

Rauenthalite, locality at Rauenthal vein system, Ste-Marie-aux-Mines, Alsace, France.

Rauvite, from composition, *Ra, U,* and *V.*

Realgar, from Arabic *Rahj al ghar* (= *powder of the mine*).

Rectorite, for E. W. Rector (1849–?), American lawyer, Hot Springs, Arkansas.

Reddingite, locality in Redding township, Fairfield County, Connecticut, where the mineral was originally found.

Redingtonite, locality at Redington mercury mine, Napa County, California.

Redledgeite, locality at Red Ledge mine, Nevada County, California.

Reedmergnerite, for Frank S. Reed (1894–) and John L. Mergner (1894–), two technicians of the U.S. Geological Survey, who prepared thin and polished sections for Survey geologists.

Reevesite, for Frank Reeves (1886–?), American geologist, who was responsible for the discovery of the Wolf Creek meteorite crater (in 1947) in Western Australia.

Refikite, for Refik Bey, for his interest in the sciences; probably the Turkish journalist (died 1865), one of the founders of the New Ottoman Society.

Reinerite, for Willy Reiner, senior chemist of the Tsumeb Corporation, South West Africa.

Renardite, for Alphonse François Renard (1842–1903), Belgian geologist and mineralogist, University of Ghent.

Renierite, for Armand Renier, Belgian geologist, director of the Geological Survey of Belgium.

Retgersite, for Jan Willem Retgers (1856–1896), Dutch chemical crystallographer.

Retzian, for Anders Jahan Retzius (1742–1821), Swedish naturalist.

Reyerite, for Eduard Reyer (1849–1914), Austrian geologist, Vienna.

Rezbanyite, locality at Rezbánya, Rumania.

Rhabdophane, from Greek for *rod* and *to appear,* in allusion to the absorption bands seen in its spectrum.

Rhodesite, for Cecil John Rhodes (1853–1902), British colonial statesman and financier, formed the DeBeers Mining Company.

Rhodium, from Greek *rhodon* (= *rose*), in allusion to the rose color of many of its salts.

Rhodizite, from Greek for *to be rose-colored,* because it tinges the blowpipe flame red.

Rhodochrosite, from Greek for *rose* and *color,* in allusion to its common color.

Rhodolite, from Greek for *rose,* in allusion to its color.

Rhodonite, from Greek for *rose,* in allusion to its characteristic color.

Rhodostannite, from Greek for *rose,* in allusion to its color, and from its chemical relationship to stannite.

Rhodusite, locality in the Eocene Flysch rocks of the island of Rhodes (Rhodus).

Rhomboclase, from its rhomboid crystal habit and Greek for *fracture,* in allusion to its perfect basal cleavage.

Rhönite, locality where it was first identified, the Rhön district of Germany.

Richellite, locality at Richelle, near Visé, Liège, Belgium.

Richetite, for Emile Richet (1884–1938), Chief Geologist, Union Minière du Haut-Katanga.

Richterite, for Theodor Richter (1824–1898), German mineralogist.

Rickardite, for Thomas Arthur Rickard (1864–1953), mining engineer, editor of *Engineering and Mining Journal,* of New York.

Riebeckite, for Emil Riebeck (1853–1885), German explorer.

Rijkeboerite, for A. Rijkeboer, chief of the analytical department, N. V. Hollandse Metallurgische Industrie Billiton, Arnhem, Netherlands.

Rilandite, for James L. Riland, newspaper publisher of Meeker, Colorado, on whose claim the mineral was discovered.

Ringwoodite, for Alfred Edward Ringwood (1930–), Australian geologist, Australian National University, Canberra.

Rinneite, for Friedrich Wilhelm Berthold Rinne (1863–1933), German crystallographer and petrographer.

Ripidolite, from Greek for *fan,* in allusion to a common mode of grouping of the crystals.

Rivadavite, for Bernardino Rivadavia (1780–1845), first President of the Argentine Republic, and promoter of mineralogical studies in that country.

Riversideite, locality at Crestmore, Riverside County, California.

Robertsite, for Willard Lincoln Roberts, American mineralogist, South Dakota School of Mines and Technology, Rapid City.

Robinsonite, for Stephen Clive Robinson (1911–), Canadian mineralogist, Queens University, Kingston, and Geological Survey of Canada, Ottawa.

Rockbridgeite, locality in Rockbridge County, Virginia.

Rodalquilarite, locality at Rodalquilar gold deposit, Almeria Province, Spain.

Roeblingite, for Washington Augustus Roebling (1837–1926), civil engineer and appreciative mineral collector, Trenton, New Jersey.

Roedderite, for Edwin Woods Roedder (1919–), U.S. Geological Survey, who first discovered the compound in synthetic materials.

Roemerite, for Friedrich Adolf Roemer (1809–1869), German geologist.

Roggianite, for Aldo G. Roggiani, Italian mineralogist.

Romanechite, locality at Romanèche, Saône-et-Loire, France.

Romarchite, from the initials of Royal Ontario Museum (ROM) and *arch*aeology, the institution where the mineral was studied.

Romeite, for Jean Baptiste Romé Delisle (1736–1770), French crystallographer.

Röntgenite, for Wilhelm Konrad Roentgen (1845–1923), German physicist, discoverer of X-rays; named because X-ray methods alone proved its existence and established its formula.

Rooseveltite, for Franklin Delano Roosevelt (1882–1945), American statesman and thirty-second President.

Roquesite, for Maurice Roques (1911–), French geologist, University of Clermont-Ferrand.

Rosasite, locality at the Rosas mine, Sulcis, Sardinia.

Roscherite, for Walter Roscher, German mineral collector of Ehrenfriedersdorf, Saxony.

Roscoelite, for Henry Enfield Roscoe (1833–1915), English chemist of Manchester, who was the first to prepare pure vanadium.

Roselite, for Gustav Rose (1798–1873), German mineralogist, University of Berlin.

Rosenbuschite, for Karl Harry F. Rosenbusch (1836–1914), German geologist and mineralogist, Heidelberg.

Rosenhahnite, for Leo Rosenhahn, American amateur mineralogist, San Anselmo, Marin County, California.

Rosickyite, for Vojtech Rosický (1880–1942), Czech mineralogist, director of the Mineralogical and Petrographical Institute of Masaryk University at Brno.

Rosieresite, locality at Rosiérès, near Carmaux, Tarn, France.

Rossite, for Clarence Samuel Ross (1880–1975), American geologist and mineralogist, U.S. Geological Survey.

Rösslerite, for Karl Rössler of Hanau, Germany.

Roubaultite, for Marcel Roubault, French geologist, University of Nancy.

Routhierite, for Pierre Routhier of Paris, France.

Roweite, for George Rowe (1868–?), English-American mineral collector of Franklin, New Jersey.

Rowlandite, for Henry Augustus Rowland (1848–1901), American physicist, Johns Hopkins University, Baltimore.

Rozenite, for Zygmunt Rozen (1874–1936), Polish mineralogist and petrographer, Academy of Mining and Metallurgy, Cracow.

Rozhkovite, for Ivan Sergeevich Rozhkov (1908–), Russian geologist, Yakutsk, Siberia, student of gold and platinum deposits.

Rubellite (var. of elbaite), from Latin *rubellus* (= *reddish*), in allusion to its color.

Ruby (var. of corundum), from Latin *rubeus* (= *red*), in allusion to its color.

Rucklidgeite, for John Christopher Rucklidge (1938–), Canadian mineralogist, University of Toronto.

Ruizite, for Joe Ana Ruiz, American amateur mineralogist and collector of Mammoth, Arizona, who discovered the mineral.

Rusakovite, for Mikhail Petrovich Rusakov (1892–), Russian geologist of Kazakhstan, Soviet Union.

Russellite, for Arthur Edward Ian Montagu Russell (1878–1964), British mineralogist.

Rustenburgite, locality at Rustenburg platinum mine in the Bushveld Igneous Complex, South Africa.

Rustumite, for Rustum Roy (1924–), materials chemist, Pennsylvania State University, State College.

Ruthenarsenite, from composition, *ruthen*ium *arsen*ide.

Rutheniridosmine, from composition, alloy of *ruthen*ium, *irid*ium, and *osmi*um.

Ruthenium, from Latin *Ruthenia,* the name for Russia, because the element was found associated with platinum in the Ural Mountains.

Ruthenosmiridium, from composition, alloy of *ruthen*ium, *osm*ium, and *iridium.*

Rutherfordine, for Ernest Rutherford (1871–1937), British atomic physicist; the mineral is radioactive.

Rutile, from Latin *rutilus* (= *red*), in allusion to the color.

Rynersonite, for the Rynerson family, San Diego County, California; members of the family have operated mines in the Himalaya pegmatite-aplite dike system since 1900.

S

Sabugalite, locality in Sabugal County in the province of Beira Alta, Portugal.

Safflorite, from German *Safflor* (= *safflower*), in allusion to its use as the pigment zaffer.

Sahamalite, for Thure Georg Sahama (1910–), Finnish geochemist, Helsinki.

Sahlinite, for Carl Andreas Sahlin (1861–?), Swedish scientist.

Sainfeldite, for Paul Sainfeld, geologist, who collected the type material in Alsace.

Sakhaite, from the name of Siberia in the Yakutian language, Soviet Union.

Sakharovaite, for M. S. Sakharova, Russian mineralogist.

Sakuraiite, for Kin-ichi Sakurai, Japanese amateur mineralogist, Tokyo.

Salammoniac (originally two words: sal ammoniac), from Medieval Latin for salt of Ammon, used by early writers for common rock salt from near the oracle of Ammon in Egypt; later the name was applied to the present compound, which was manufactured in Egypt.

Saleeite, for Achille Salée (1883–1932), Belgian professor, Louvain.

Salesite, for Reno H. Sales (1876–   ), American geologist, Anaconda Copper Mining Company.

Salmonsite, for Frank A. Salmons (1866–?), of Pala, California.

Samarskite, for Vasilii Erafovich Samarski-Bykhovets (1803–1870), of the Russian Corps of Mining Engineers.

Sampleite, for Mat Sample, Chile Exploration Company, Chuquicamata, Chile.

Samsonite, locality at the Samson vein, Andreasberg, Harz Mountains, Germany.

Samuelsonite, for Peter B. Samuelson, American prospector, Rumney, New Hampshire.

Sanbornite, for Frank Sanborn, American mineralogist, Division of Mines, Department of Natural Resources, California.

Sanderite, for Bruno Sander (1884–   ), Austrian geologist, Alte Universität, Innsbruck.

Sanidine, from Greek for *tablet* or *board*, in allusion to the typical tabular habit of the mineral.

Sanjuanite, locality in Department of Pocito, San Juan province, Argentina.

Sanmartinite, for José de San Martín (1778–1850), liberator of Argentina.

Santafeite, for the Atchison, Topeka and Santa Fe Railroad Company, in recognition of its pioneer exploration and development of the uranium deposits of New Mexico.

Santanaite, locality at Mina Santa Ana, Caracoles, Sierra Gorda, Chile.

Santite, for Giorgi Santi (1746–1823), Tuscan naturalist and director of the Museum of Natural History, Pisa, Italy.

Saponite, from Latin *sapo* (= *soap*), on account of its soaplike appearance.

Sapphire (var. of corundum), an ancient term of uncertain origin, possibly related to Hebraic *sappir* and Sanskrit *sanipruja;* applied by the ancients, including the Greeks, to lazurite.

Sapphirine, from *sapphire,* on account of its common sapphire-blue color.

Sarabauite, locality at Sarabau mine, Sarawak, Malaysia.

Sarcolite, from Greek for *flesh*, in allusion to the flesh-red color.

Sarcopside, from Greek for *flesh* and *to view*, in allusion to the fleshlike color.

Sard (var. of quartz), from Sardis, capitol of the ancient kingdom of Lydia.

Sarkinite, from Greek for *made of flesh*, in allusion to the blood-red color and greasy luster.

Sarmientite, for Domingo Faustino Sarmiento (1811–1888), Argentinian educator and statesman.

Sartorite, for Sartorius von Waltershausen (1809–1876), who first reported the species.

Saryarkite, from the Kazakh name of the steppes of central Kazakhstan, Soviet Union.

Sassolite, locality at Sasso, Tuscany, Italy, where it was first found.

Satimolite, locality which apparently is in Kazakhstan, Soviet Union.

Satpaevite, for Kanysh Imantaevich Satpaev (1899–1964), Soviet geologist, Kazakhstan.

Sauconite, locality in Upper Saucon, Lehigh County, Pennsylvania.

Sazhinite, for Nikolai Petrovich Sazhin (1898–1969), founder of the Soviet rare-earth industry.

Sborgite, for Umberto Sborgi (1883–1955), Italian chemist.

Scacchite, for Arcangelo Scacchi (1810–1894), Italian mineralogist, University of Naples.

Scapolite (mineral group name), from Greek for *shaft*, in allusion to the common stumpy prismatic habit.

Scarbroite, locality at South Bay, Scarborough, Yorkshire, England.

Scawtite, locality at Scawt Hill, County Antrim, Ireland.

Schachnerite, for Doris Schachner (1904–), German mineralogist, Technische Hochschule, Aachen.

Schafarzikite, for Ferenc Schafarzik (1854–1927), Hungarian mineralogist.

Schairerite, for John Frank Schairer (1904–1970), American physical chemist, Geophysical Laboratory, Washington, D.C.

Schallerite, for Waldemar Theodore Schaller (1882–1967), American mineralogist, U.S. Geological Survey.

Schaurteite, for Werner T. Schaurte, German scientist.

Scheelite, for Karl Wilhelm Scheele (1742–1786), Swedish chemist.

Schertelite, for Arnulf Schertel (1841–1902), professor, Freiberg Mining Academy, Germany.

Scheteligite, for Jacob Schetelig (1875–1935), Norwegian mineralogist, director of the Mineralogical Museum, Oslo.

Schirmerite, for J. F. L. Schirmer, Superintendent of the United States Mint at Denver, Colorado.

Schmiederite, apparently for Oscar Schmieder (1891–), German geographer who spent time at Córdoba, Argentina.

Schmitterite, for Eduardo Schmitter Villada, Mexican mineralogist and petrologist, National University of Mexico.

Schneiderhöhnite, for Hans Schneiderhöhn (1887–1962), German mineralogist, University of Freiburg.

Schoderite, for William Paul Schoder (1900–), American research chemist, Union Carbide Nuclear Company.

Schoenfliesite, for Artur Moritz Schönflies (1853–1928), German mathematician, University of Frankfurt, who studied crystal symmetry and derived the 230 space groups.

Schoepite, for Alfred Schoep (1881–1966), Belgian mineralogist, who contributed much to the mineralogy of uranium.

Scholzite, for Dr. A. Scholz, German collector and chemist, Regensburg.

Schoonerite, for Richard Schooner, American mineral collector and student of New England minerals, of Woodstock, Connecticut.

Schorl, from German *Schörl*, a term with a long and uncertain history; perhaps from the locality Schörlau (meaning Schörl village) in Germany, however, the village may have been named for the mineral. Perhaps the name was derived from the Old German word *Schor*, meaning *refuse* or *impurities*. Originally the name was applied to a large number of minerals.

Schorlomite, from mineral schorl and Greek *homos* (= *the same*), in allusion to its resemblance to schorl.

Schreibersite, for Karl Franz Anton von Schreibers (1775–1852), Vienna, Austria.

Schreyerite, for Werner Schreyer (1930–), professor, Ruhr University, Bochum, Germany.

Schroeckingerite, for J. von Schroeckinger, Austrian geologist, who found and described the mineral's occurrence at Joachimsthal, Bohemia.

Schubnelite, for H. J. Schubnel, French mineralogist, Bureau de Recherches Géologiques et Minières.

Schuetteite, for Curt Nicolaus Schuette (1895–), American mining engineer and geologist, a specialist on mercury deposits.

Schuilingite, for H. J. Schuiling, chief geologist for Union Minière du Haut-Katanga.

Schultenite, for August Benjamin de Schulten (1856–1912), who prepared the synthetic compound.

Schwartzembergite, for a Dr. Schwartzemberg, an assayer at Copiapó, Chile, who first drew attention to the mineral.

Scolecite, from Greek for *worm*, because the borax bead of the mineral sometimes curls up like a worm.

Scorodite, from Greek for *garlic*, in allusion to the odor of the mineral obtained on heating.

Scorzalite, for Evaristo Pena Scorza (1899–), Brazilian mineralogist.

Seamanite, for Arthur Edmund Seaman (1858–1937), American geologist and mineralogist, Michigan College of Mining and Technology, Houghton, Michigan.

Searlesite, for John W. Searles, California pioneer.

Sederholmite, for Jakob Johannes Sederholm (1863–1934), Finnish geologist, director of the Geological Survey of Finland.

Sedovite, for Georgii Iakovlevich Sedov (1877–1914), Russian polar investigator.

Seeligerite, for Erich Seeliger, German mineralogist, Technical University, Berlin.

Segelerite, for Curt George Segeler (1901–), American amateur mineralogist, Brooklyn, New York.

Seidozerite, locality in the Seidozero region, Lovozero massif, Kola Peninsula, Soviet Union.

Seinäjokite, locality at Seinäjoki deposit, central Finland.

Sekaninaite, for Josef Sekanina (1901–), Czech mineralogist, who first found the mineral in 1928.

Selenite (var. of gypsum), from Greek for *moon*, in allusion to the moon-like white reflections from the mineral.

Selenium, from Greek for *moon*, so named from its being associated with tellurium, named from Latin *tellus* (= *the Earth*).

Selenolite, from composition, *selen*ious *o*xide.

Selen-tellurium, from composition, *selen*ium and *tellurium*.

Seligmannite, for Gustav Seligmann (1849–1920), German mineral collector, Koblenz.

Sellaite, for Quintino Sella (1827–1884), Italian mining engineer and mineralogist.

Semenovite, for E. I. Semenov, Russian mineralogist, Institute of Mineralogy and Geochemistry of Rare Elements, Moscow.

Semseyite, for Andor von Semsey (1833–1923), Hungarian nobleman interested in minerals and meteorites.

Senaite, for Joachim da Costa Sena, professor, Ouro Preto, Brazil.

Senarmontite, for Henri Hureau de Sénarmont (1808–1862), French physicist and mineralogist, School of Mines, Paris, who first described the species.

Senegalite, locality in Senegal, the country in which the mineral was first found.

Sengierite, for Edgar Sengier (1879–1963), an official of the Union Minière du Haut-Katanga.

Sepiolite, from Greek for *cuttle-fish,* the bone of which is light and porous like the mineral.

Septechlorite (mineral group name), from Latin *septem* (= *seven*), and the mineral name chlorite, in allusion to the seven angstrom spacing of its structural layers and its chemical relationship to chlorite.

Serandite, for J. M. Sérand, West African mineral collector, who assisted in its collection on the island of Rouma (Republic of Guinea).

Serendibite, from *Serendib,* an old Arabic name for Ceylon.

Sericite (var. of mica, usually muscovite), from Greek word for *silky,* in allusion to its silky luster.

Serpentine (mineral group name), from Latin *serpens* (= *snake*), in allusion to the surface pattern of some serpentinite rocks, which resemble the skin of a serpent.

Serpierite, for Giambattista Serpieri (1832–1897), Italian engineer, active in the development of the mines at Laurium, Greece.

Shadlunite, for Tatyana Shadlun, Russian mineralogist, well-known for her research on ore minerals.

Shandite, for Samuel James Shand (1882–1957), Scottish-American petrologist, Columbia University, New York.

Sharpite, for R. R. Sharp, who discovered the uranium deposit at Shinkolobwe, Katanga district, Belgian Congo, in 1915.

Shattuckite, for the Shattuck Arizona Copper Company's mine at Bisbee, Arizona.

Shcherbakovite, for Dmitrii Ivanovich Shcherbakov (1893–1966), Russian mineralogist and geochemist.

Shcherbinaite, for V. V. Shcherbina, Russian geochemist.

Sheridanite, locality in Sheridan County, Wyoming.

Sherwoodite, for Alexander M. Sherwood (1888–   ), American analytical chemist, U.S. Geological Survey.

Shortite, for Maxwell Naylor Short (1889–1952), American mineralogist, University of Arizona.

Sibirskite, apparently for locality in Siberia, Soviet Union.

Sicklerite, for the Sickler family of Pala, California, near where the mineral was discovered.

Siderazot, from composition, iron (*sideros*) and nitrogen (*azote,* from Greek for *without life*).

Siderite, from composition, an iron (*sideros*) mineral.

Sideronatrite, from composition, iron (*sideros*) and sodium (*natrium*).

Siderophyllite, from composition, iron (*sideros*), and Greek for *leaf,* because of its foliated structure.

Siderotil, from composition, iron (*sideros*), and Greek for *fiber,* in allusion to its fibrous to needlelike structure.

Siegenite, locality in the Siegen district of Westphalia, Germany.

Sigloite, locality at the Siglo XX mine, Llallagua, Bolivia.

Silhydrite, from composition, a *silica hydrate.*

Sillenite, for Lars Gunnar Sillén (1916–?), of Stockholm, Sweden, who did much work on the bismuth oxides.

Sillimanite, for Benjamin Silliman (1779–1864), American chemist and geologist, Yale University.

Silver, an Old English word for the metal, earlier spelled *seolfer;* related to Dutch *zilver* and German *Silber.*

Simonellite, for Vittorio Simonelli (1860–?), Italian geologist, who first observed the mineral.

Simplotite, for J. R. Simplot, of the J. R. Simplot Mining Company, Boise, Idaho.

Simpsonite, for Edward Sydney Simpson (1875–1939), government mineralogist and analyst, of Western Australia.

Sincosite, locality near Sincos, Department of Junin, Peru.

Sinhalite, from *Sinhala,* the old Sanskrit name for Ceylon, a locality for the mineral.

Sinnerite, for Rudolf von Sinner, president of the Commission of the Naturhistorisches Museum, Bern, Switzerland.

Sinoite, from composition, *Si, N,* and *O.*

Sjögrenite, for Hjalmar Sjögren (1856–1922), Swedish mineralogist, University of Stockholm.

Skinnerite, for Brian John Skinner (1928–   ), American geochemist, Yale University, who studied the synthetic compound.

Sklodowskite, for Marie Sklodowska Curie (1867–1934), Polish-French physicist and chemist, a pioneer in the study of radioactivity.

Skutterudite, locality at Skutterud, Norway.

Slavikite, for František Slavík (1876–1957), Czech mineralogist, Charles University, Prague.

Slavyanskite, locality at the Slavyansk salt cupola, southeastern Dneprovsk-Donets basin, Soviet Union.

Smaltite, from *smalt,* deep blue cobalt oxide pigment; the mineral is a source of the pigment.

Smectite, from Greek for *fuller's earth;* one of the clays included under the general name fuller's earth.

Smithite, for George Frederick Herbert Smith (1872–?), English gemologist and crystallographer, of the British Museum.

Smithsonite, for James Smithson (1765–1829), English chemist and mineralogist, whose endowment founded the Smithsonian Institution in Washington, D.C.

Smolianinovite, for Nikolai Alekseevich Smolianinov (1885–?), Russian mineralogist.

Smythite, for Charles Henry Smyth, Jr. (1866–1937), American economic geologist and petrologist, Princeton, New Jersey.

Sobolevskite, for Petr Grigorevich Sobolevski (1781–1841), Russian metallurgist who studied the platinum deposits of the Urals.

Sobotkite, locality at Mt. Sobotka in Lower Silesia, Poland.

Soda alum, from composition, the sodium member of the alum group.

Sodalite, from composition, it contains sodium (soda).

Soda niter, from composition, like niter, but with predominant sodium.

Soddyite, for Frederick Soddy (1877–1956), British radiochemist and physicist.

Sodium autunite, from composition, like autunite, but with predominant sodium.

Sodium betpakdalite, from composition, chemically similar to betpakdalite, but with predominant sodium.

Sodium boltwoodite, from composition, like boltwoodite, but with predominant sodium.

Sodium uranospinite, from composition, chemically similar to uranospinite and meta-uranospinite, but with predominant sodium.

Sodium-zippeite, from composition, like zippeite, but with predominant sodium.

Sogdianite, from Sogdiana, the name of an ancient state of Middle Asia.

Söhngeite, for G. Söhnge, chief geologist of the Tsumeb Corporation, South West Africa.

Solongoite, locality at the Solongo contact-metamorphic boron deposit, Buryat A.S.S.R., Soviet Union.

Sonolite, locality at the Sono manganese mine, Japan.

Sonoraite, locality at the Moctezuma mine in Sonora, Mexico.

Sorbyite, for Henry Clifton Sorby (1826–1908), English chemist, the founder of metallography.

Sorensenite, for Henning Sorensen (1926–　), Danish geologist, University of Copenhagen.

Souzalite, for Antonio José Alves de Souza, director of the Departmento nacional da produção mineral, Brazil.

Spadaite, for Lavino Spada de Medici (1801–1863), Italian political figure and mineralogist.

Spangolite, for Norman Spang (1842–?), of Etna, Allegheny County, Pennsylvania, who supplied the original specimen for study.

Spar (obsolete term), from Old English *spaerstan* (*spar stone*), apparently originally for gypsum; later used for many minerals which are vitreous, crystalline, and easily cleavable, e.g., heavy spar (barite), pearl spar (dolomite), calcspar (calcite), fluorspar (fluorite), feldspar, and many others.

Spencerite, for Leonard James Spencer (1870–1959), English mineralogist and editor, Keeper of Minerals in the British Museum of Natural History, London.

Sperrylite, for Francis Louis Sperry (?–1906), chemist of Sudbury, Ontario, who first found the mineral.

Spessartine, locality at Spessart in northwestern Bavaria.

Sphaerocobaltite, from Greek for *ball* and the element cobalt, in allusion to the spheroidal shapes and composition.

Sphalerite, from Greek for *treacherous* or *slippery*, the mineral was often mistaken for galena, but yielded no lead.

Sphene (syn. for titanite), from Greek for *wedge*, in allusion to the characteristic habit of the crystals.

Spinel, apparently from the Latin *spinella* (= *little thorn*), in allusion to the spine-shaped octahedral crystals.

Spiroffite, for Kiril Spiroff (1901–　), American mineralogist, Michigan College of Mining and Technology, Houghton.

Spodiosite, from Greek for *ash gray*, in allusion to its color.

Spodumene, from Greek for *reduced to ashes*, either in allusion to its ash-gray color, or the ash-colored mass formed when heated before the blowpipe.

Spurrite, for Josiah Edward Spurr (1870–1950), American geologist.

Stanfieldite, for Stanley Field (1875–1964), chairman of the board of trustees of the Field Museum of Natural History, Chicago.

Stannite, from composition, a tin (*stann*um) mineral.

Stannoidite, from the mineral stannite, in allusion to its similar composition.

Stannopalladinite, from composition, tin (*stann*um) and *palladi*um.

Staringite, for Winand Carel Hugo Staring (1808–1877), a founder of geology and mineralogy in the Netherlands.

Starkeyite, locality at the Starkey mine, Madison County, Missouri.

Staurolite, from Greek for *cross*, in allusion to its common cruciform twins.

Steenstrupine, for Knud Johannes Vogelius Steenstrup (1842–1913), Danish geologist of Copenhagen.

Steigerite, for George Steiger (1869–1944), American chemist, U.S. Geological Survey.

Stellerite, for Georg Wilhelm Steller (1709–1746), German explorer and zoologist.

Stenhuggarite, from Swedish *stenhuggar* (= *stonemason*), for Brian Harold Mason (1917–), New Zealand-American geochemist and mineralogist, U.S. National Museum, Washington, D.C., who has contributed much to the study of Långban, Sweden, minerals.

Stenonite, for Nicolaus Steno, the Latinized name of Niels Steensen (1638–1686), Danish anatomist and discoverer of the law of the constancy of interfacial angles on crystals.

Stepanovite, for Pavl Ivanovich Stepanov (1880–1947), Russian geologist.

Stephanite, for Victor Stephan (1817–1867), Archduke of Austria and mining director.

Stercorite, from Latin *stercus* (= *dung*), in reference to the occurrence of the mineral in guano deposits.

Sternbergite, for Count Kaspar Maria Sternberg (1761–1838), Czech botanist, who helped establish the National Museum in Prague.

Sterryite, for Thomas Sterry Hunt (1826–1892), first mineralogist with the Geological Survey of Canada.

Stetefeldtite, for Carl August Stetefeldt (1838–1896), German-American mining engineer and metallurgist.

Stevensite, apparently for the Stevens Institute of Technology, Hoboken, New Jersey.

Stewartite, locality at the Stewart mine, Pala, San Diego County, California.

Stibarsen, from composition, antimony (*stib*ium) and *arsen*ic.

Stibiconite, from composition, antimony (*stib*ium), and Greek for *powder* or *dust*, because it often occurs as a powder.

Stibiocolumbite, from composition, antimony (*stib*ium) and niobium (*columbi*um).

Stibiopalladinite, from composition, antimony (*stib*ium) and *palladi*um.

Stibiotantalite, from composition, antimony (*stib*ium) and *tantal*um.

Stibnite, from Greek *stimmi* and Latin *stibi,* old names applied to the mineral.

Stichtite, for Robert Sticht, general manager of the Mt. Lyell Company's mining properties, Tasmania.

Stilbite, from Greek for *to shine,* in allusion to the pearly to vitreous luster.

Stilleite, for Hans Stille (1876–1966), German geologist.

Stillwaterite, locality in the Stillwater Complex, Montana.

Stillwellite, for Frank Leslie Stillwell (1888–?), Australian mineralogist.

Stilpnomelane, from Greek for *shining* and *black,* in allusion to its luster and color.

Stishovite, for Sergei Mikhailovich Stishov, Russian crystallographer, Moscow, who first synthesized the substance.

Stistaite, from composition, antimony (*sti*bium) and tin (*sta*nnum).

Stokesite, for George Gabriel Stokes (1819–1903), British mathematician and physicist, Cambridge University.

Stolzite, for Joseph Alexis Stolz (1803–1896), of Teplitz, Bohemia, who first drew attention to the species.

Stottite, for Charles E. Stott, geologist, director of the Tsumeb mine, South West Africa.

Stranskiite, for Iwan N. Stranski (1897–), German physicist and chemist, Berlin.

Strashimirite, for Strashimir Dimitrov, Bulgarian petrographer.

Strätlingite, for W. Strätling, who synthesized the compound in 1938.

Strelkinite, for M. F. Strelkin (1905–1965), Russian mineralogist, a student of uranium ores.

Strengite, for Johann August Streng (1830–1897), German mineralogist, University of Giessen.

Stringhamite, for Bronson Ferrin Stringham (1907–1968), American mineralogist, University of Utah, Salt Lake City.

Stromeyerite, for Friedrich Stromeyer (1776–1835), German chemist, Göttingen, who first analyzed the mineral.

Strontianite, locality at Strontian in Argyllshire, Scotland.

Strontioborite, from composition, a *stronti*um *bor*ate.

Strontiodresserite, from composition, *stronti*um, and the similarity of its formula to dresserite.

Strontioginorite, from composition, like ginorite, but with predominant strontium.

Strontium-apatite, from composition, a strontium-rich member of the apatite group.

Strunzite, for Hugo Strunz (1910–), German mineralogist, Technische Universität, Berlin.

Strüverite, for Giovanni Strüver (1842–1915), Italian mineralogist, University of Rome.

Struvite, for Heinrich Christian Gottfried von Struve (1772–1851), geologist-mineralogist of the Russian diplomatic service.

Studtite, for Franz Edward Studt, Belgian geologist.

Stumpflite, for E. F. Stumpfl, professor, Mining Institute, Leoben, Austria.

Sturtite, for Charles Sturt (1795–1869), explorer and first white man to visit Broken Hill, New South Wales, Australia, where the mineral occurs.

Stützite, for Andreas Stütz (1747–1806), Austrian mineralogist, Vienna.

Suanite, locality at Suan County, Korea.

Sudburyite, locality at Sudbury, Ontario, Canada.

Sudoite, for Toshio Sudo (1911–), Japanese mineralogist and crystallographer, University of Tokyo.

Sugilite, for Ken-ichi Sugi (1901–1948), Japanese petrologist, who first found the mineral.

Sukulaite, locality at Sukula, Tammela, in southwestern Finland.

Sulfoborite, from composition, contains *sulf*ate and *bor*ate.

Sulfur, from Latin *sulfur*, an old name for the element; akin to Sanskrit *sulvere*.

Sulphohalite, from composition, a *sulf*ate with the *hal*ogen elements Cl and F.

Sulvanite, from composition, contains *sul*fur and *van*adium.

Suolunite, locality in Inner Mongolia gave its name to the mineral.

Surinamite, locality in Surinam, the country in which the mineral was first found.

Sursassite, from Sursass, the Roman name for Oberhalbstein, Switzerland.

Susannite, locality at the Susanna mine, Leadhills, Scotland.

Sussexite, locality in Sussex County, New Jersey.

Svabite, for Anton Svab (1703–1768), Swedish mining official.

Svanbergite, for Lars Fredrik Svanberg (1805–1878), professor, University of Uppsala, Sweden.

Svetlozarite, for Svetlozar Ionchev Botev (1940–1971), Bulgarian zeolite mineralogist.

Swartzite, for Charles Kephart Swartz (1861–1949), American mineralogist and geologist, Johns Hopkins University, Baltimore, Maryland.

Swedenborgite, for Emanuel Swedenborg (1688–1772), Swedish religious philosopher, mathematician, and engineer.

Swinefordite, for Ada Swineford (1917–   ), American geologist and clay mineralogist, Western Washington State College, Bellingham.

Switzerite, for George Shirley Switzer (1915–   ), American mineralogist, Museum of Natural History, U.S. National Museum, Washington, D.C.

Sylvanite, from Tran*sylvan*ia (Rumania), where it was first found, and in allusion to the element tellurium (*sylvanium* is old name proposed for the element), which it contains.

Sylvite, from the old chemical name for the substance, *Sal digestivus Sylvii* or digestive salt of Francois Sylvius de le Boë (1614–1672), Dutch physician and chemist of Leyden.

Symplesite, from Greek words meaning *to bring together*, in allusion to its relationship to other rare arsenate minerals.

Synadelphite, from Greek for *with* and *brother*, in allusion to the fact it occurs with several chemically similar minerals.

Synchysite, from Greek word for *confounding*, in allusion to its being originally mistaken for parisite.

Syngenite, from Greek word meaning *related*, because of its chemical resemblance to polyhalite.

Szaibelyite, for Stephan Szaibely (1777–1855), Hungarian mine surveyor of Rézbánya, who first collected the mineral.

Szmikite, for Ignaz Szmik, Hungarian mining official at Felsöbánya, Rumania.

Szomolnokite, locality at Szomolnok, Czechoslovakia.

T

Taaffeite, for Edward Charles Richard Taaffe (1898–   ), Bohemian-Irish gemologist, of Dublin.

Tacharanite, from Gaelic *tacharan* (= *a changeling*), because on standing in air the mineral breaks down to form other compounds.

Tachyhydrite, from Greek for *quick* and *water*, in allusion to its ready deliquescence.

Tadzhikite, locality in Tadzhik S.S.R., central Asia, Soviet Union.

Taeniolite, from Greek for *band* or *strip*, in allusion to the form of the crystals, which is tabular.

Taenite, from Greek for *band* or *strip*, in allusion to its thin platy structure.

Takanelite, for Katsutoshi Takane (1899–1945), Japanese mineralogist, professor, Tohoku University.

Takovite, locality in Takovo, Serbia, Yugoslavia.

Talc, of ancient origin, probably derived from Arabic *talq.*

Talnakhite, locality in Talnakh ore deposit, Norilsk, western Siberia, Soviet Union.

Tamarugite, locality in the Tamarugal Pampa, Chile.

Tantalaeschynite, from composition, like aeschynite, but with predominant tantalum.

Tantalite, from Tantalus of Greek mythology; named in allusion to the tantalizing difficulties encountered in making a solution of the mineral in acids preparatory to analysis.

Tanteuxenite, from composition, like euxenite, but with predominant tantalum.

Tanzanite (var. of zoisite), locality in Tanzania, Africa.

Tapiolite, from Tapio, in Finnish mythology a god of the forest; mineral was discovered in Finland.

Taramellite, for Torquato Taramelli (1845–1922), Italian geologist.

Taramite, locality in Wali-tarama Valley, near Mariupol, Ukraine, Soviet Union.

Taranakite, locality at Sugarloaves, Taranaki, New Zealand.

Tarapacaite, locality in Tarapacá province, Chile.

Tarasovite, for Taras Grigorievich Shevchenko (1814–1861), Ukrainian writer.

Tarbuttite, for Percy Coventry Tarbutt, director of the Broken Hill Exploration Company, Rhodesia, who collected some of the first specimens.

Tatarskite, for Vitaly Borisovich Tatarskii (1907–), Russian mineralogist and geologist, Leningrad University.

Tavorite, for Elysairio Tavora (1911–), Brazilian mineralogist, University of Brazil, Rio de Janeiro.

Taylorite, for William J. Taylor (1833–1864), mineral chemist, Philadelphia, Pennsylvania, and Maryland.

Tazheranite, locality in the Tazheran alkalic massif, west of Lake Baikal, Siberia, Soviet Union.

Teallite, for Jethro Justinian Harris Teall (1849–1924), Director General of the Geological Survey of Great Britain and Ireland.

Teepleite, for John Edgar Teeple (1874–1931), American chemist, in recognition of his contributions to the chemistry of Searles Lake, California.

Teineite, locality at Teine mine, Hokkaido, Japan.

Telargpalite, from composition, *tel*lurium, silver (*arg*entum), and *pal*ladium.

Tellurantimony, from composition, *tellu*rium and *antimony.*

Tellurite, from composition, a *tellur*ium dioxide.

Tellurium, from Latin *tellus* (= *the Earth*).

Tellurobismuthite, from composition, *tellur*ium and *bismuth*.

Temagamite, locality at Temagami copper deposit, Ontario.

Tengerite, for C. Tenger, Swedish chemist, who early studied the mineral.

Tennantite, for Smithson Tennant (1761–1815), English chemist.

Tenorite, for Michele Tenore (1781–1861), Italian botanist, University of Naples.

Tephroite, from the Greek word for *ash-colored*, in allusion to its color.

Terlinguaite, locality at Terlingua, Brewster County, Texas.

Ternovskite, locality at Ternovsky iron mine, Krivoy-Rog, Kherson, Soviet Union.

Tertschite, for Hermann Tertsch (1880–1962), Austrian mineralogist and crystallographer, University of Vienna.

Teruggite, for Mario E. Teruggi, geologist, Universidad Nacional La Plata, Argentina.

Teschemacherite, for Frederick Edward Teschemacher (1791–1863), English chemist, who first described the species.

Testibiopalladite, from composition, *te*llurium, antimony (*stibi*um), and *pallad*ium.

Tetradymite, from Greek for *four* and *twin*, in allusion to the twinned crystals, which are often fourlings.

Tetraferroplatinum, from Greek for *four*, in allusion to the tetragonal structure, and from composition, iron (*ferr*um) and *platinum*.

Tetrahedrite, from the tetrahedral form of the crystals.

Tetrakalsilite, from Greek for *four*, and from composition, K, Al, Si; a polymorph of kalsilite which has its a axis four times longer.

Tetranatrolite, from crystal system, the *tetra*gonal dimorph of natrolite.

Tetrawickmanite, from crystal system, the *tetra*gonal dimorph of wickmanite.

Texasite, locality in rare-earth rich pegmatites in the state of Texas.

Thalcusite, from composition, *thal*lium, Cu, and S.

Thalenite, for Tobias Robert Thalén (1827–1905), Swedish physicist.

Thaumasite, from Greek for *astonishment*, in allusion to its remarkable composition which contains silicate, carbonate, and sulfate components.

Thenardite, for Louis Jacques Thénard (1777–1857), French chemist, University of Paris.

Thermonatrite, from Greek word for *heat* and from natron, because it forms from the drying of natron when it is heated.

Thomsenolite, Hans Peter Jörgen Julius Thomsen (1826–1909), Danish physical chemist, founder of the Greenland cryolite industry, who first noticed the species.

Thomsonite, for Thomas Thomson (1773–1852), Scottish chemist, who analyzed the mineral.

Thorbastnaesite, from composition, like bastnaesite, but with predominant thorium.

Thoreaulite, for Jacques Thoreau, professor, University of Liège, Belgium.

Thorianite, from composition, *thor*ium dioxide.

Thorite, from composition, contains *thor*ium.

Thorogummite, from composition, *thor*ium, and Latin *gummi,* for *gum,* in allusion to its gumlike appearance.

Thorosteenstrupine, from composition, like steenstrupine, but with predominant thorium.

Thortveitite, for Olaus Thortveit, Norwegian engineer of Iceland.

Thorutite, from composition, *thor*ium, *U,* and *Ti.*

Thucholite (mixture of uraninite with carbonaceous matter), from composition, *Th, U, C, H,* and *O.*

Thulite (var. of zoisite), from Thule, an ancient name of Norway.

Thuringite (var. of chamosite), locality near Saalfeld, in Thuringia, Germany.

Tiemannite, for Johann Carl Wilhelm Ferdinand Tiemann (1848–1899), German chemist of Berlin, who discovered the mineral.

Tienshanite, locality in Turkestan-Alai alkalic province, southern Tienshan, Soviet Union.

Tikhonenkovite, for Igor Petrovich Tikhonenkov (1927–1961), Russian student of alkalic rocks and minerals.

Tilasite, for Daniel Tilas (1712–1772), Swedish mining engineer.

Tilleyite, for Cecil Edgar Tilley (1894–1973), British petrologist, Cambridge University.

Tin, an Old English word for the metal; related to Dutch *tin* and German *Zinn.*

Tinaksite, from composition, *Ti, Na, K,* and *Si.*

Tincalconite, from Sanskrit *tincal* (= *borax*), and from Greek for *powder,* in allusion to the fact it can form from the dehydration of borax and has a pulverulent form.

Tinticite, locality near Tintic Standard mine, Tintic District, Utah.

Tintinaite, locality at the Tintina silver mines, Yukon, Canada.

Tinzenite, locality at Tinzen, Grisons, Switzerland.

Tirodite, locality at Tirodi, Central Provinces, India.

Titanite (syn. for sphene), from composition, contains titanium.

Tlalocite, from Tlaloc, the ancient Mexican god of rain, in allusion to the high content of water.

Tlapallite, from the Nahua (Mexican Indian) word *tlapalli*, for *paint*, in allusion to the appearance of the mineral as paintlike films on rock fracture surfaces at the type locality.

Tobermorite, locality near Tobermory, Island of Mull, Scotland.

Tochilinite, for Mitrofan Stepanovich Tochilin, professor at Voronezh University, Voronezh region, Soviet Union.

Tocornalite, for S. F. Tocornal, rector of Santiago University, Chile.

Todorokite, locality at the Todoroki mine, Hokkaido, Japan.

Tombarthite, for Thomas Fredrik Weiby Barth (1899–1971), Norwegian geochemist, Oslo University.

Topaz, from Greek *Topazion*, the name of an island in the Red Sea, meaning *to seek*, the island often being covered by mist.

Topazolite (var. of andradite), from topaz, in allusion to its similar yellow color and luster.

Torbernite, for Torbern Olof Bergman (1735–1784), Swedish chemist and mineralogist.

Törnebohmite, for Alfred Elis Törnebohm (1838–1911), Swedish geologist, pioneer in the study of Archean rocks of central Sweden.

Torreyite, for John Torrey (1796–1873), American naturalist and chemist, who studied Franklin, New Jersey, minerals as early as 1822.

Tosudite, for Toshio Sudo (1911–   ), Japanese mineralogist and crystallographer, University of Tokyo.

Tourmaline (mineral group name), from Singhalese *turamali*, a term originally applied to zircon and other gems by jewelers of Ceylon.

Tranquillityite, locality in the Sea of Tranquillity on the Moon, from which the Apollo XI rocks were collected.

Traskite, for John Boardman Trask (1824–1879), American geologist, first State Geologist of California.

Treasurite, locality at Treasure Vault lode, Park County, Colorado.

Trechmannite, for Charles O. Trechmann (1851–1917), English crystallographer.

Tremolite, locality in Tremola Valley, near St. Gotthard, Switzerland.

Trevorite, for Tudor Gruffydd Trevor (1865–1958), Welsh-South African geologist, mining inspector for Pretoria district, Transvaal.

Tridymite, from Greek for *three* and *twin*, in allusion to the twinned crystals, which are often trillings.

Trigonite, from Greek for *triangle*, in reference to the trigonal shape of the crystals.

Trikalsilite, from Greek for *three*, and from composition, *K, Al, Si*; a polymorph of kalsilite which has its *a* axis three times longer.

Trimerite, from Greek for *in three parts*, in allusion to its unusual optical structure.

Triphylite, from Greek for *three* and *family*, in allusion to its three cations, iron, lithium, and manganese.

Triplite, from Greek or Latin for *threefold*, probably in allusion to its three cleavages.

Triploidite, from triplite, in allusion to its resemblance to triplite in physical characteristics and composition.

Trippkeite, for Paul Trippke (1851–1880), Polish mineralogist, who discovered the mineral.

Tripuhyite, locality at Tripuhy near Ouro Preto, Minas Geraes, Brazil.

Tritomite, from Greek for *three* and *to cut*, alluding to the trihedral cavities which the acute rhombohedral crystals left imprinted on the gangue mineral.

Trögerite, for a Mr. R. Tröger, a mining official at Schneeberg, Saxony.

Trogtalite, locality at Trogtal quarry near Lautenthal, Harz Mountains, Germany.

Troilite, for Domenico Troili, Italian natural scientist, who first noted its occurrence in a meteorite in 1766.

Trolleite, for Hans Gabriel Trolle-Wachtmeister (1782–1871), Swedish chemist.

Trona, from the Arabic name of the native salt.

Troostite (var. of willemite), for Gerard Troost (1776–1850), Dutch-American mineralogist, State Geologist of Tennessee.

Truscottite, for Samuel John Truscott (1870–?), English mining geologist.

Trüstedtite, for Otto Trüstedt, whose work on prospecting methods led to the discovery of the Outokumpu deposit, Finland.

Tschermakite, for Gustav Tschermak von Seysenegg (1836–1927), Austrian mineralogist, Vienna.

Tschermigite, locality at Tschermig, Bohemia, Czechoslovakia.

Tsumcorite, for the Tsumeb Corporation, Tsumeb mine, South West Africa.

Tsumebite, locality at Tsumeb, South West Africa.

Tucekite, for Karel Tuček, Czech mineralogist, curator at the National Museum in Prague.

Tugtupite, locality at Tugtup agtakorfia, Greenland.

Tuhualite, locality at Opo Bay, Mayor (Tuhua) Island, New Zealand.

Tulameenite, locality in Tulameen River area, British Columbia, Canada.

Tundrite, locality in the Lovozero tundra, Soviet Union.

Tunellite, for George Tunell (1900–   ), American geochemist, University of California, Los Angeles.

Tungstenite, from composition, *tungsten* disulfide.

Tungstite, from composition, a *tungsten* oxide mineral.

Tungusite, locality near Tunguska River, Siberia, Soviet Union.

Tunisite, locality in Tunisia, the country in which the mineral was first found.

Turanite, locality in the Turan region of Turkestan, Soviet Union.

Turgite (hematite with adsorbed water), locality near Turja River near Bogoslovsk, Urals, Soviet Union.

Turquoise, from French word meaning *Turkish;* the original stones came into Europe from the Persian locality through Turkey.

Tuscanite, locality in Tuscany region of Italy.

Tvalchrelidzeite, for A. A. Tvalchrelidze, founder of the Georgian mineralogical-petrographic school, Soviet Union.

Tveitite, for John Tveit, Norwegian quarry owner of Hoydalen, Telemark, Norway, who discovered the mineral.

Twinnite, for Robert Mitchell Thompson (1918–1967), Canadian mineralogist, University of British Columbia; the name Thompson means ''son of Thomas,'' the later being Aramaic for *twin.*

Tychite, from Greek for *luck* or *chance,* in allusion to the circumstance that of two crystals found in a stock of about 5000 northupite crystals the first and one of the last ten proved to be tychite.

Tyretskite, locality near Tyretsk station of the East Siberian railway, Soviet Union.

Tyrolite, locality in Tyrol, Austria.

Tyrrellite, for Joseph Burr Tyrrell (1858–1957), Canadian geologist, who first visited area where the mineral was discovered.

Tysonite, for S. T. Tyson, who collected and supplied the specimens used in the original study of the mineral.

Tyuyamunite, from the locality at the hill Tyuya Muyun, in Ferghana, Turkestan, Soviet Union; the hill, traversed by the gorge of the Aravan River, resembles a camel's hump.

U

Uduminelite, locality at Udumine, Fukushima Prefecture, Japan.

Ugrandite (subgroup of the garnet group), from the initial parts of *u*varovite, *gro*ssular, and *and*radite.

Uhligite, for Alfred Louis Johannes Uhlig (1883–1919), German geologist, leader of the African expedition on which the material was first collected.

Uklonskovite, for Aleksandr Sergeievich Uklonskii (1888–?), Soviet mineralogist, Academy of Sciences, Uzbek S.S.R., Soviet Union.

Ulexite, for George Ludwig Ulex (1811–1883), German chemist, who discovered the mineral.

Ullmannite, for Johann Christoph Ullmann (1771–1821), German (Hessian) chemist and mineralogist, who first discovered the mineral.

Ulvöspinel, locality at Södra Ulvön, Angermanland, northern Sweden, and the fact it is in the spinel group.

Umangite, locality at Sierra de Umango, Argentina.

Umbozerite, locality near Umbozera in the Lovozero alkalic massif, Soviet Union.

Umohoite, from composition, *U, Mo, H,* and *O.*

Ungemachite, for Henri-Léon Ungemach (1880–1936), Belgian crystallographer.

Uralborite, locality in the Urals, Soviet Union, and from composition, a borate.

Uralite (pseudomorph of amphibole after pyroxene), locality in the Ural Mountains, Soviet Union, where it was first noticed.

Uralolite, locality in the Ural region, Soviet Union.

Uramphite, from composition, *ur*anyl *am*monium *ph*osphate.

Uraninite, from composition, in allusion to its uranium content.

Uranite, from composition, in allusion to its uranium content.

Uranocircite, from composition, uranium, and Greek for *falcon,* for locality at Falkenstein, Saxony.

Uranophane, from composition, uranium, and Greek for *to appear,* in allusion to the presence of the element.

Uranopilite, from composition, uranium, and Greek for *felt,* in reference to its velvety appearance.

Uranospathite, from composition, uranium, and Greek word for *a broad blade,* in allusion to the crystal habit.

Uranosphaerite, from composition, uranium, and Greek for *sphere,* in allusion to its globular forms.

Uranospinite, from composition, uranium, and Greek word for *siskin* (*finch*), in allusion to the siskin-green color.

Urea, from its occurrence in urine.

Ureyite, for Harold Clayton Urey (1893–   ), American chemist and Nobel Laureate.

Uricite, from uric, in allusion to its occurrence in urine.

Urvantsevite, for Nikolai Nikolaevich Urvantsev (1893–), Russian geologist and Polar investigator of Leningrad, for research in the Noril'sk deposits.

Usovite, for Mikhail Antonovich Usov (1883–1939), Soviet geologist.

Ussingite, for Niels Viggo Ussing (1864–1911), professor, Copenhagen, Denmark.

Ustarasite, locality at Ustarasaisk, western Tien-Shan, Soviet Union.

Uvanite, from composition, *u*ranium and *van*adium.

Uvarovite, for Count Sergei Semeonovich Uvarov (1786–1855), Russian nobleman, Imperial Academy of St. Petersburg.

Uvite, locality in province of Uva of Ceylon (Sri Lanka).

V

Vaesite, for Johannes Vaes, mineralogist for the Union Minière du Haut-Katanga, Belgian Congo.

Valentinite, for Basilius Valentinus (pseudonym for Johannes Thölde), German alchemist who worked in the late sixteenth and early seventeenth centuries, and wrote on the properties of antimony.

Vallerite, for Göran Vallerius (1683–1744), Swedish mining geologist.

Vanadinite, from composition, in allusion to its vanadium content.

Vanalite, from composition, *van*adium and *al*uminum.

Vandenbrandeite, for Pierre van den Brande (1896–1957), Belgian geologist, who discovered the ore deposit at Kalongwe, Katanga district, Belgian Congo.

Vandendriesscheite, for Adrien Vandendriessche (1914–1940), Belgian mineralogist and geologist.

Vanoxite, from composition, *van*adium *ox*ide.

Vanthoffite, for Jacobus Hendricus van 'tHoff (1852–1911), Dutch physical chemist.

Vanuralite, from composition, *van*adium, *ur*anium, and *al*uminum.

Vanuranylite, from composition, *van*adium and *uranyl.*

Variscite, from Latin *Variscia,* ancient name of the Voigtland district in Germany where the mineral was first found.

Varlamoffite, for Nicolas Varlamoff, mining engineer, Belgian Congo, who discovered the mineral.

Varulite, locality at Varuträsk, Sweden.

Vashegyite, locality at Vashegy near Szirk, Comitat Gömör, Hungary.

Vaterite, for Heinrich Vater, German mineralogist and chemist, professor, Tharandt, Saxony.

Vauquelinite, for Louis Nicolas Vauquelin (1763–1829), French chemist and discoverer of chromium.

Vauxite, for George Vaux, Jr. (1863–1927), American mineral collector of Bryn Mawr, Pennsylvania.

Väyrynenite, for Heikki Allan Väyrynen (1888–?), Finnish geologist.

Veatchite, for John A. Veatch, the first to detect the presence of borates in the mineral water of California.

Veenite, for R. W. van der Veen, metallographer.

Vermiculite, from Latin *vermiculare* (= *to breed worms*), in allusion to the peculiar exfoliation exhibited when specimens are rapidly heated.

Vernadite, for Vladimir Ivanovich Vernadskii (1863–1945), Russian naturalist and geochemist.

Verplanckite, for William E. Ver Planck (1916–1963), American geologist, California Division of Mines and Geology.

Vertumnite, from Vertumnus, the mighty Etruscan god; venerated by the ancient people who lived in the region of Italy where the mineral was discovered.

Vesignieite, for Louis Vésignié (1870–1954), French mineral collector, president of the Mineralogical Society of France in 1932.

Vesuvianite, locality at Mt. Vesuvius, Italy, where it was found in ejected blocks.

Veszelyite, for A. Veszelyi, Hungarian mining engineer, who discovered the mineral.

Vikingite, for the vikings, who were early explorers of Greenland, the country where the mineral was first discovered.

Villamaninite, locality near Villamanin, León Province, Spain.

Villiaumite, for a Mr. Villiaume, French explorer, in whose collection of rocks from Guinea the mineral was first found.

Vimsite, for the All-Union Research Institute of Mineral Resources, Soviet Union; the initials of the Russian name are VIMS.

Vincentite, for Ewart Albert Vincent (1919–   ), English geochemist, Oxford University.

Vinogradovite, for Aleksander Pavlovich Vinogradov (1895–1975), Russian geochemist.

Violarite, from Latin *violaris* (= *of violet*), in allusion to its color.

Virgilite, for Virgil Everett Barnes (1903–   ), American geologist, University of Texas at Austin, in honor of his pioneering work in the study of tektites and other natural glasses.

Viseite, locality at Visé, a town in Belgium.

Vishnevite, locality in Vishneviye mountains, Urals, Soviet Union.

Vivianite, for J. G. Vivian, English mineralogist, who discovered the mineral.

Vladimirite, locality near Vladimirskoe, Altai Mountains, Soviet Union.

Vlasovite, for Kuzma Aleksevich Vlasov (1905–1964), Russian mineralogist and geochemist, who has worked in the Lovozero massif.

Voglite, for Josef Florian Vogl, Austrian mining official and mineralogist, who published on the uranium minerals of Joachimsthal, Bohemia.

Volborthite, for Alexander von Volborth (1800–1876), Russian paleontologist, who first noticed the mineral.

Volkovskite, for A. I. Volkovskaya, Soviet petrographer, who first found the mineral.

Voltaite, for Alessandro Giuseppe Antonio Anastasio Volta (1745–1827), Italian physicist.

Volynskite, for I. S. Volynskii (1900–1962), Soviet mineralogist, Institute of Mineralogy, Geochemistry, and Crystal Chemistry of Rare Elements, Moscow.

Vonsenite, for Magnus Vonsen (1879–1954), American mineral collector of Petaluma, California, who was interested in borate minerals.

Vrbaite, for Karl Vrba (1845–1922), Bohemian mineralogist.

Vuagnatite, for Marc Bernard Vuagnat (1922–   ), Swiss geologist, University of Geneva, who has studied ophiolite rocks.

Vulcanite, locality at Vulcan, Gunnison County, Colorado.

Vuonnemite, locality in the valley of the Vuonnemi River, Khibina, Kola Peninsula, Soviet Union.

Vysotskite, for N. K. Vysotskii, Soviet geologist, who first found the platinum metals at Noril'sk.

W

Wad (general term for soft manganese oxides), from a provincial English word for black, soft powders; origin not known.

Wadeite, for Arthur Wade, Australian geologist, who collected the mineral.

Wagnerite, for F. M. von Wagner (1768–1851), German mining official in Munich.

Wairakite, locality at Wairakei, in the central part of the North Island, New Zealand.

Wairauite, locality in Wairau Valley, South Island, New Zealand.

Wakabayashilite, for Yaichiro Wakabayashi (1874–1943), Japanese mineralogist, Mitsubishi Mining Company, Japan.

Wakefieldite, locality near Wakefield, Quebec, Canada.

Wallisite, presumably for locality, Valais (Wallis) Canton, Switzerland.

Walpurgite, locality in Walpurgis vein of the Weisser Hirsch mine at Neustädtl, Schneeberg, Saxony.

Walstromite, for Robert E. Walstrom, American mineral collector, Fresno, California, who discovered the mineral.

Wardite, for Henry Augustus Ward (1834–1906), American naturalist, collector, and dealer, Rochester, New York.

Wardsmithite, for Ward Conwell Smith (1906–   ), American geologist, U.S. Geological Survey, Menlo Park, California.

Warwickite, locality near Warwick, Orange County, New York.

Water, an Old English word for the substance; related to Dutch *water,* German *Wasser,* and Russian *voda.*

Wattevillite, for Oscar de Watteville, of Paris, France.

Wavellite, for William Wavell (?–1829), English physician, Horwood Parish, Devonshire, who discovered the mineral.

Waylandite, for Edward James Wayland, first director of the Uganda Geological Survey.

Weberite, for Theobald Weber (1823–1886), one of the founders of the cryolite industry in Denmark.

Weddellite, locality at Weddell Sea, Antarctica.

Weeksite, for Alice Mary Dowse Weeks (1909–   ), American mineralogist, Temple University, Philadelphia.

Wegscheiderite, for Rudolf Wegscheider (1859–?), chemist, who formed the compound synthetically.

Wehrlite, for Alois Wehrle (1791–1835), Austrian mineralogist and chemist.

Weibullite, for Kristian Oskar Mats Weibull (1856–1923), who first described the mineral.

Weilerite, locality at Weiler bei Lahr, Schwarzwald, Germany.

Weilite, for René Weil (1901–   ), French mineralogist of Strasbourg, known for his studies of Alsatian minerals.

Weinschenkite (syn. for churchite), for Ernst Weinschenk (1865–1921), German petrographer, University of Munich.

Weissbergite, for Bryon G. Weissberg, New Zealand geologist-mineralogist, Department of Scientific and Industrial Research, Petone.

Weissite, for Louis Weiss, owner of the Good Hope mine, Gunnison County, Colorado.

Welinite, for Eric Welin, mineralogist and geochronologist.

Wellsite, for Horace Lemuel Wells (1855–1924), American chemist, Yale University.

Weloganite, for William Edmond Logan (1798–1875), Canadian geologist, first director of the Geological Survey of Canada.

Welshite, for Wilfred R. Welsh, amateur mineralogist and teacher of the natural sciences, Upper Saddle River, New Jersey.

Wenkite, for Eduard Wenk (1907–   ), Swiss mineralogist and petrologist, University of Basel.

Wermlandite, locality at Långban, Wermland (Värmland), Sweden.

Wernerite (syn. for scapolite group), for Abraham Gottlob Werner (1750–1817), German mineralogist and geologist, who separated geology from mineralogy.

Westerveldite, for Jan Westerveld (1905–1962), Dutch mineralogist and economic geologist, University of Amsterdam.

Westgrenite, for Arne Fredrik Westgren (1889–1975), Swedish X-ray crystallographer who made the synthetic compound.

Wherryite, for Edgar Theodore Wherry (1885–?), American botanist and mineralogist, U.S. National Museum and University of Pennsylvania.

Whewellite, for William Whewell (1794–1866), English natural scientist and philosopher.

Whitlockite, for Herbert Percy Whitlock (1868–1948), American mineralogist, curator of minerals at the American Museum of Natural History, New York.

Whitmoreite, for Robert Whitmore, American amateur mineral collector, Weare, New Hampshire, who collected the mineral.

Wickenburgite, locality near Wickenburg, Arizona.

Wickmanite, for Franz-Erik Wickman (1915–   ), Swedish-American mineralogist, who made many contributions to the study of Långban, Sweden, minerals.

Widenmannite, for Johann Friederich Wilhelm Widenmann (1764–1798), German mineralogist, who first discovered uranium micas in the Black Forest.

Wightmanite, for R. H. Wightman, Director of Exploration and Mining, Riverside Cement Company, California.

Wilkeite, for R. M. Wilke, American mineral collector and dealer, Palo Alto, California.

Wilkmanite, for Wanold Wrydon Wilkman (1872–?), Finnish geologist.

Willemite, for William I (Willem Frederik) (1772–1843), king of the Netherlands from 1815 to 1840.

Willemseite, for Johannes Willemse (1909–   ), geologist, University of Pretoria, South Africa.

Williamsite (var. of antigorite, a serpentine), for Lewis White Williams (1804?–1873), of West Chester, Chester County, Pennsylvania.

Willyamite, locality at Broken Hill mines, Willyama township, New South Wales, Australia.

Winchite, for H. J. Winch, who discovered it.

Wiserite, for David Friedrich Wiser (1802–1878), Swiss mineralogist and chemist, who first analyzed it.

Witherite, for William Withering (1741–1799), English physician, botanist and mineralogist.

Wittichenite, locality near Wittichen, Black Forest, Baden, Germany.

Wittite, for Th. Witt, Swedish mining engineer.

Wodginite, locality at Wodgina, Australia.

Wöhlerite, for Friedrich Wöhler (1800–1882), German chemist and professor, Göttingen.

Wolfeite, for Caleb Wroe Wolfe (1908–   ), American crystallographer, Boston University.

Wolframite, origin uncertain, perhaps from German *Wolf* (= *wolf*), and *Rahm* (= *froth*), in allusion to an objectionable scum formed during the smelting of tin ores containing tungsten; or perhaps from German *Wolfrig,* or related terms, used by early Saxon miners, in allusion to the action of the mineral, when present in tin ores, to decrease (eat away) the yield of tin during concentration or smelting.

Wolframoixiolite, from composition, tungsten (*wolfram*), and the fact the mineral is similar to both wolframite and ixiolite in structure.

Wollastonite, for William Hyde Wollaston (1766–1828), English mineralogist and chemist.

Wölsendorfite, locality at Wölsendorf, Bavaria.

Woodhouseite, for Charles Douglas Woodhouse (1888–1975), American mineral collector of Santa Barbara, California, founder of California Federation of Mineral Societies.

Woodruffite, for Samuel Woodruff, American miner and mineral collector who specialized in specimens from Franklin, New Jersey.

Woodwardite, for Samuel Pickworth Woodward (1821–1865), English naturalist and geologist.

Wroewolfeite, for Caleb Wroe Wolfe (1908–   ), American crystallographer, Boston University.

Wulfenite, for Franz Xaver Wülfen [Wulffen] (1728–1805), Austrian Jesuit and mineralogist, who wrote a monograph on the lead ores of Carinthia, southern Austria.

Wurtzite, for Charles Adolphe Wurtz (1817–1884), French chemist, professor of organic chemistry, Paris.

Wüstite, for Ewald Wüst, German geologist.

Wyartite, for Jean Wyart (1902–   ), French mineralogist and crystallographer, Sorbonne, Paris.

Wyllieite, for Peter John Wyllie (1930–   ), British-American petrologist and geochemist, University of Chicago.

X

Xanthiosite, from Greek for *yellow* and *sulfur*, in allusion to the color of the mineral.

Xanthoconite, from Greek for *yellow* and *powder*, in allusion to the color of its streak.

Xanthophyllite (syn. for clintonite), from Greek for *yellow* and *leaf*, in allusion to its color and foliated structure.

Xanthoxenite, from Greek for *yellow* and *guest* or *stranger*, named for the original supposed chemical resemblance to cacoxenite (meaning *bad guest*).

Xenotime, from Greek for *stranger* and *to honor*, in reference to the fact the crystals are small and rare, and were long unnoticed; it was originally misspelled as kenotime, as if from Greek for *vain* and *to honor*.

Xingzhongite (= hsingchungite), for Chinese type locality, no specific name given.

Xocomecatlite, from the Nahua (Mexican Indian) word for *grapes*, in allusion to the clusters of green spherules.

Xonotlite, locality at Tetela de Xonotla, Mexico.

Y

Yagiite, for Kenzo Yagi, Japanese geologist, Hokkaido University, Sapporo.

Yaroslavite, locality by that name in Siberia, Soviet Union.

Yavapaiite, for the Yavapai Indian tribe that inhabited the area of Arizona, near Jerome, where the mineral occurs.

Yeatmanite, for Pope Yeatman (1861–1953), American mining engineer, associated with mining at Franklin, New Jersey.

Yedlinite, for Neal Yedlin (1908–1977), American mineral collector and micromounter, New Haven, Connecticut, who first observed the species.

Yftisite, from composition, *Y, F, Ti,* and *Si.*

Yixunite (= yihsünite), for Chinese type locality, no specific name given.

Yoderite, for Hatten Schuyler Yoder, Jr. (1921–), American petrologist, director of the Geophysical Laboratory, Washington, D.C.

Yofortierite, for Yves Oscar Fortier (1914–), Canadian geologist, director of the Geological Survey of Canada.

Yoshimuraite, for Toyofumi Yoshimura, Japanese mineralogist, professor, Kyushu University.

Yttrialite, from composition, the yttrium rare earths being the chief cations.

Yttrocolumbite, from composition, yttrium, and its relationship to columbite and yttrotantalite.

Yttrocrasite, from composition, yttrium, and Greek for *mixture*, because with yttrium there are many other elements.

Yttrotantalite, from composition, *yttr*ium and *tantal*um.

Yttrotungstite, from composition, *yttr*ium and *tungst*en, and similarity to cerotungstite.

Yugawaralite, locality at Yugawara hot springs, Kanagawa Prefecture, Japan.

Yukonite, locality at Tagish Lake, Yukon Territory, Canada.

Yuksporite, locality at Yuksporlack in the Kola Peninsula, Soviet Union.

Z

Zaherite, for Mohammad Abduz Zaher, Geological Survey of Bangladesh, who discovered the mineral.

Zairite, locality at Eta-Etu, Kivu, Zaire.

Zapatalite, for Emiliano Zapata (1879–1919), a popular hero of the Mexican Revolution.

Zaratite, for a Señor Zarate, of nineteenth-century Spain.

Zavaritskite, for Aleksandr Nikolaevich Zavaritskii (1884–1952), Russian petrographer.

Zektzerite, for Jack Zektzer, of Seattle, Washington, who initiated the study of the mineral.

Zellerite, for Howard D. Zeller, American geologist, U.S. Geological Survey, Denver, Colorado, discoverer of the mineral.

Zemannite, for Josef Zemann (1923–), Austrian crystallographer, University of Vienna, who contributed much to the understanding of the structure of tellurium compounds.

Zeolite (mineral group name), from Greek for *to boil*, because of its intumescence before the blowpipe flame.

Zeophyllite, from Greek for *to boil* and *leaf,* probably in allusion to the half-spherical forms which have a foliated structure.

Zeunerite, for Gustav Anton Zeuner (1828–1907), German physicist, director of the School of Mines at Freiberg, Saxony.

Zhemchuzhnikovite, for Yuri Appollonovich Zhemchuzhnikov (1885–1957), Russian geologist and clay mineralogist.

Zinalsite, from composition, *zinc, Al,* and *Si.*

Zinc, from German *Zink,* of obscure origin.

Zincaluminite, from composition, *zinc* and *alumin*um, and its similarity to aluminite.

Zincite, from composition, *zinc* is the major metal.

Zinc-melanterite, from composition, like melanterite, but with predominant zinc.

Zincobotryogen, from composition, like botryogen, but with predominant zinc.

Zincocopiapite, from composition, like copiapite, but with predominant zinc.

Zincrosasite, from composition, like rosasite, but with predominant zinc.

Zincsilite, from composition, a *zinc sil*icate.

Zinc-zippeite, from composition, like zippeite, but with predominant zinc.

Zinkenite, for J. K. L. Zinken (1798–1862), mineralogist and mining geologist.

Zinnwaldite, locality at Zinnwald, Bohemia, so named because of the tin (German *Zinn*) veins found there.

Zippeite, for Franz Xaver Maximilian Zippe (1791–1863), Austrian mineralogist who first studied the mineral.

Zircon, from Arabic *zarqun,* which was derived from Persian *zar* (= *gold*), and *gun* (= *color*).

Zircophyllite, from composition, contains *zirco*nium, and its relationship to astro*phyllite.*

Zircosulfate, from composition, a hydrated *zirco*nium *sulfate.*

Zirkelite, for Ferdinand Zirkel (1838–1912), German petrographer, University of Leipzig.

Zirklerite, for a Dr. Zirkler, director of the Aschersleben Potash Works, Germany.

Zirsinalite, from composition, *zir*conium, *Si,* and *Na.*

Zoisite, for Siegmund Zois, Baron von Edelstein (1747–1819), Austrian scholar and writer who was interested in minerals and who financed mineral-collecting expeditions.

Zorite, from Russian expression referring to "the rosy radiance of the sky at dawn," in allusion to the rosy color.

Zunyite, locality at the Zuni mine, near Silverton, San Juan County, Colorado.

Zussmanite, for Jack Zussman (1924–), crystallographer and mineralogist, Manchester University, England.

Zvyagintsevite, for Orest Evgenevich Zvyagintsev (1894–1967), Russian geochemist, who did research on the platinum metals.

Zwieselite, locality at Rabenstein, near Zwiesel, Bavaria.

Zykaite, for Václav Zýka, Czech geologist, director of the Institute of Mineral Raw Materials, Kutná Hora.

Appendix I
Mineral Names Whose
Derivations Were Not Found

Balavinskite
Iriginite
Slawsonite
Velikite

Appendix II
Personal Names Not Readily Discernible in Mineral Names

Aldrin, Edwin E., arm*al*colite
Anelli, Franco, franco*anelli*te
Armstrong, Neil A., *arm*alcolite
Axon, H. J., h*axon*ite
Barnes, Virgil E., *virgil*ite
Barth, Thomas F. W., tom*barth*ite
Belov, Nikolai V., arsenate-*belov*ite
Bergman, Torbern O., *torbern*ite
Borneman-Starynkevich, Irina D., *borneman*ite
Botev, Svetlozar I., *svetlozar*ite
Bravo, Eliodoro Bellido, *bellido*ite
Brooks, Alfred Hulse, *huls*ite
Cathcart, Charles M. (Lord *Greenock*), *greenock*ite
Collins, Michael, arm*alcol*ite
Conti, Piero Ginori, *ginori*te
deFeuffe, J. S. P. J. Delvaux, *delvaux*ite
de la Peyrous, Picot, *picot*ite
de Saint-Fond, Barthélemy Faujas, *faujas*ite
de Villiers, A. J. F. M. Brochant, *brochant*ite
Dickson, Frank W., frank*dickson*ite
Dimitrov, Strashimir, *strashimir*ite
Donnay, Gabrielle, gai*donnay*ite
Ekanayake, F. L. D., *ekan*ite
Ericson, Leif, *leif*ite
Field, Stanley, stan*field*ite

Fisher, D. Jerome, djer*fisher*ite
Fortier, Yves O., yo*fortier*ite
Foshag, William F., *fosh*allasite
Fries, Carl, Jr., carl*fries*ite
Frondel, Clifford, *clifford*ite
Geier, Bruno H., bruno*geier*ite
Geijer, Per A., *per*ite
Goldsmith, Julian R., jul*gold*ite
Gower, John A., ja*gower*ite
Hageman, Gustav A., *gustav*ite
Hunt, Thomas Sterry, *sterry*ite
Ito, Jun, jun*ito*ite
Joy, Laura R., *laur*ite
Key, Charles Locke, lud*lock*ite
Kingsbury, Arthur W. G., *arthur*ite
Kirchheimer, Franz W., meta*kirchheimer*ite
Kolovrat-Chervinsky, L. S., *kolovrat*ite
Larsen, Esper S., Jr., *esper*ite
Lisboa, Miguel Arrojado, *arrojad*ite
Logan, William E., we*logan*ite
Low, Albert P., ap*low*ite
Melon, Joseph, *melon*josephite
Mergner, John L., reed*mergner*ite
Mason, Brian H., *brian*ite and *stenhuggar*ite
Meyer, André M., andre*meyer*ite
Moore, Paul Brian, paul*moore*ite
Peacock, Martin A., *pavo*nite
Picot, Paul, *picot*paulite
Reed, Frank S., *reed*mergnerite
Riva, Carol, cupro*riva*ite
Rogers, Austin F., *austin*ite
Roy, Della M., *della*ite
Roy, Rustum, *rustum*ite
Russell, Arthur, *arthur*ite
Sharp, Montroyd, *montroyd*ite
Shevchenko, Taras G., *taras*ovite
Smith, Frederick Ludlow, III, *ludl*ockite
Smith, John Lawrence, *lawrenc*ite
Smith, Joseph V., joe*smith*ite
Smith, Ward C., ward*smith*ite

Strippelmann, Leo, *leo*nite
Sudo, Toshio, to*sud*ite
Termier, Henri F. E., henri*termier*ite
Thompson, James B., Jr., jim*thompson*ite
Thompson, Robert M., *twin*nite
Trolle-Wachtmeister, Hans G., *trolle*ite
Urban, Joseph J., j*urban*ite
Van Wambeke, (Mrs.) L., eylettersite
Villada, Eduardo Schmitter, *schmitter*ite
vom Rath, Maria Rosa, *maria*lite
von Semsey, Andor, *andor*ite
von Waltershausen, Sartorius, *sartor*ite
Watanabe, Manjiro, *manjiro*ite
Williams, Alpheus F., af*will*ite
Wolfe, Caleb Wroe, wroe*wolfe*ite
Zambonini, Ferruccio, *ferrucci*te

Appendix III
Glossary

acicular. Needlelike or spiny.

acronym. A word formed from the initial letters or syllables of other words; a kind of portmanteau word; illustrated by mineral names formed from standard chemical symbols, for example, sinoite from Si, N, O.

blowpipe analyses. Various chemical tests obtained largely at high temperatures by the use of a blowpipe; the blowpipe is usually a conical tube, fitted with a mouthpiece, through which air is blown to obtain increased temperatures from an ordinary gas flame.

botryoidal structure. Closely united spherical masses which resemble a bunch of grapes.

capillary structure. Hairlike or threadlike crystals.

cleavage. The property of many crystalline substances to split along planes of weakness determined by their internal crystal structure; cleavage planes are always parallel to possible crystal faces.

crystal. A substance having a specific internal structure and enclosed by symmetrically arranged planar surfaces (faces).

crystallography. The science of crystals; concerned with the causes, nature, and consequences of the periodic interatomic arrangements in solid matter.

deliquescence. The process whereby substances melt gradually and become liquid by attracting and absorbing moisture from the air.

dimorphism. The propensity shown by some substances to crystallize in two distinctly different structures; a kind of polymorphism.

efflorescence. A thin crust or coating which is often powdery.

element, chemical. One of a class of substances that consists solely of atoms of the same atomic number; illustrated by gold, copper, oxygen, silicon, and aluminum.

element, native. Term applied to chemical elements which occur free or uncombined in nature.

etymology. The history of a particular word, with all its changes of form, phonetics, spelling, and meaning; a branch of philology which treats the derivation of words.

exfoliate. To separate and come off in scales, especially when the mineral is heated.

foliated structure. In leaves or plates which separate easily.

form, crystal. A group of planar faces which have a like position with respect to the symmetry elements of the crystal; in the idealized crystal all the faces in a form are of the same shape and size; examples include cube, octahedron, tetrahedron, and rhombohedron.

globular structure. Spherical in shape.

group, mineral. Minerals showing a similarity of crystallography and structure as well as chemistry; for example, the feldspar group, the mica group, and the garnet group.

habit, crystal. The external shape of a crystal which results from the interaction of its internal structure and the environment in which it forms; examples are bladed, columnar, octahedral, prismatic, tabular, and tetrahedral.

hardness. The resistance offered by a mineral to abrasion or scratching.

intumescence. The bubbling up of a molten mass, especially when heated with a blowpipe.

isomorphic. Refers to two or more minerals which have identical or similar structures and also which show continuous variations in composition between them; for example, the plagioclase feldspars.

isostructural. Those situations where minerals have identical structures, but do not necessarily have compositional similarities; for example, halite and galena.

isotypic. Minerals which have similarities in their crystal structures and chemical compositions, but which do not necessarily have continuous variations in composition; for example calcite and siderite.

lamellar structure. Composed of thin layers or scales.

luster. The appearance of a fresh surface of a mineral in reflected light.

mineral. A natural occurring, inorganically formed, substance which has a characteristic chemical composition and usually an ordered atomic arrangement, which is ofttimes expressed in external geometrical forms.

mineralogy. The science of minerals; especially concerned with the minerals of the Earth, Moon, and meteorites.

mnemonic. Pertaining to, aiding, or designed to aid the memory.

optic axes. Those directions in noncubic crystals along which there is no double refraction. There is one optic axis (uniaxial) in the hexagonal and tetragonal crystal systems, and two optic axes (biaxial) in the orthorhombic, monoclinic, and triclinic systems.

petrology. The science of rocks, including the study of rocks by all available methods; concerned with rock origins, present conditions, classifications, alterations, etc.

pleochroism. Variation in the color of a single crystal resulting from the differential absorption of white light (usually polarized) in different crystallographic directions.

plumose structure. Feathery appearance or structure.

polymorphism. The ability of some substances with one chemical makeup to crystallize in more than one structural arrangement; for example the different structures of carbon represented by diamond, graphite, chaoite, and lonsdaleite.

polytypism. A kind of polymorphism in which the different structural arrangements differ slightly by the manner in which identical unit layers of structure are stacked upon each other; stacking variations give rise to different unit cell dimensions and may produce different space groups and crystal systems.

portmanteau word. A word created by combining parts of existing words; for example biopyribole from *bio*tite, *py*roxene, and amph*ibole*.

pseudomorphism. The phenomenon resulting when a mineral is replaced by another mineral without any change in the external shape.

reniform structure. Large, rounded, kidney-shaped masses.

rock. A naturally formed aggregate composed of one or more minerals, whether or not coherent, constituting an essential or appreciable part of the Earth or other celestial bodies.

series, mineral. Applied to minerals related by isomorphism, in which the crystal structures are essentially the same, and there are continuous variations in composition between them; for example, the plagioclase feldspar series.

species, mineral. A distinct individual mineral with a characteristic composition and crystal structure.

specific gravity. The ratio of the weight of a mineral to the weight of an equal volume of water at $4°C$.

streak. Color of the fine powder of the mineral.

system, crystal. Based on their symmetry characteristics all crystals can be assigned to one of six large groups referred to as systems; these include isometric (or cubic), hexagonal, tetragonal, orthorhombic, monoclinic, and triclinic.

twinning. Those situations where two or more crystals of the same species are grown together in a nonparallel, but a symmetrical and rational fashion.

variety, mineral. A variant of a mineral species, usually resulting from either small amounts of chemical substitution (impurities), or because of a distinctive habit, color, or other physical property.

General Bibliography

Aballain, M., Chambolle, P., Derec-Poussier, F., Guillemin, C., Mignon, R., Pierrot, R., and Sarcia, J. A. *Index Alphabetique de Nomenclature Mineralogique*. Paris: Bureau de Recherches Geologiques et Minieres, 1968.

Aikin, Arthur, and Aikin, Charles R. *Dictionary of Chemistry and Mineralogy*, 3 Volumes. London: J. and A. Arch, 1807–1814.

Allan, Thomas. *Alphabetical List of the Names of Minerals*. Edinburgh: Canterbury Press, 1808.

Allan, Thomas. *Mineralogical Nomenclature, Alphabetically Arranged; with Synoptic Tables of the Chemical Analyses of Minerals*. Edinburgh: A. Constable & Co., 1814.

Arem, Joel E. *Color Encyclopedia of Gemstones*. New York: Van Nostrand Reinhold Co., 1977.

Bailey, Dorothy, and Bailey, Kenneth C. *An Etymological Dictionary of Chemistry and Mineralogy*. London: E. Arnold & Co., 1929.

Bertrand, Élie. *Dictionnaire Universel des Fossiles Propres*, 2 Volumes. La Haye: P. Gosse, Jr. & D. Pinet, 1763.

Bradley, John E. S., and Barnes, Archibald C. *Chinese-English Glossary of Mineral Names*. New York: Consultants Bureau, 1963.

Bristow, Henry W. *A Glossary of Mineralogy*. London: Longman, Green, Longman & Roberts, 1861.

Chambers's Mineralogical Dictionary, New Ed. Brooklyn: Chemical Publishing Co., 1954.

Chester, Albert H. *A Dictionary of the Names of Minerals*. New York: John Wiley & Sons, 1896.

Chester, Albert H. *A Catalogue of Minerals, Alphabetically Arranged, with Their Chemical Composition and Synonyms*, Third Ed. New York: John Wiley & Sons, 1897.

Dana, Edward S. *System of Mineralogy of Dana*, Sixth Ed. New York: John Wiley & Sons, 1892.

Dana, Edward S. *First Appendix to the Sixth Edition of Dana's System of Mineralogy.* New York: John Wiley & Sons, 1899.

Dana, Edward S., and Ford, W. E. *Second Appendix to the Sixth Edition of Dana's System of Mineralogy.* New York: John Wiley & Sons, 1909.

Dana, James D. *System of Mineralogy,* Fifth Ed. New York: John Wiley & Son, 1868.

Debus, Allen G. (editor). *World Who's Who in Science.* Chicago: Marquis-Who's Who, Inc., 1968.

De Gallitzin, Dimitri. *Recueil de Noms par Ordre alphabetique apropriés en Minéralogie.* Brunswick: 1801.

De Michele, Vincenzo. *Dizionario di Mineralogia.* Novara: Istituto geografico De Agostini, 1970.

Egleston, Thomas. *A Catalogue of Minerals and Synonyms,* Second Ed. New York: John Wiley & Sons, 1892.

Ehlers, Curt. *Nomina der Kristallographie und Mineralogie.* Hamburg: Boysen-Maasch, 1952.

English, George L. *Descriptive List of the New Minerals, 1892–1938.* New York: McGraw-Hill, 1939.

Fleischer, Michael. *1975 Glossary of Mineral Species.* Bowie, Maryland: Mineralogical Record, Inc., 1975.

Ford, W. E. *Third Appendix to the Sixth Edition of Dana's System of Mineralogy.* New York: John Wiley & Sons, 1915.

Francke, H. Hugo A. *Über die Mineralogische Nomenclature.* Berlin: R. Friedländer & Sohn, 1890.

Frondel, Clifford. *System of Mineralogy of Dana,* Seventh Ed., Volume III, Silica Minerals. New York: John Wiley & Sons, 1962.

Gary, Margaret, McAfee, Robert Jr., and Wolf, Carol L. (editors). *Glossary of Geology.* Washington, D.C.: American Geological Institute, 1972.

Hey, Max H. *An Index of Mineral Species and Varieties Arranged Chemically,* Second Ed. London: British Museum (Natural History), 1962, 1975.

Hey, Max H. *Appendix to the Second Edition of an Index of Mineral Species and Varieties Arranged Chemically.* London: British Museum (Natural History), 1963.

Hey, Max H. *A Second Appendix to the Second Edition of an Index of Mineral Species and Varieties Arranged Chemically.* London: British Museum (Natural History), 1974.

Himus, Godfrey W. *A Dictionary of Geology.* London: Penguin Books, 1954.

Humble, William. *Dictionary of Geology and Mineralogy*, Third Ed. London: R. Griffith & Co., 1860.

Keferstein, Christian. *Mineralogia Polyglotta*. Halle: E. Anton, 1849.

Kipfer, Alex. *Mineralindex*. Thun: Ott, 1974.

Landrin, Henri. *Dictionnaire de Minéralogie, de Géologie, et de Métallurgie*. Paris: F. Didot frères, 1852.

Lüschen, Hans. *Die Namen der Steine. Das Mineralreich im Spiegel der Sprache*. Thun, Munich: Ott, 1968.

Nakhimzhan, Oskar E. *Polyglot Dictionary of Mineral Species and Varieties*. Moscow: 1962.

Page, David. *Handbook of Geological Terms, Geology and Physical Geography*, Second Ed. Edinburgh, London: W. Blackwood & Sons, 1865.

Palache, C., Berman, H., and Frondel, C. *System of Mineralogy of Dana*, Seventh Ed., Volumes I–II. New York: John Wiley & Sons, 1944, 1951.

Povarennykh, A. S. *Crystal Chemical Classification of Minerals*, 2 Volumes. New York: Plenum Press, 1972.

Readwin, T. Allison. *Index to Mineralogy*. London: E. & F. N. Spon, 1867.

Rice, Clara M. *Dictionary of Geological Terms*. Ann Arbor: Edward Brothers, 1955.

Roberts, W. L., Rapp, G. R. Jr., and Weber, J. *Encylopedia of Minerals*. New York: Van Nostrand Reinhold Co., 1974.

Schmidt, Carl W. *Geologisch-mineralogisches Wörterbuch*. Leipzig, Berlin: B. G. Teubner, 1921.

Schmidt, Carl W. *Wörterbuch der Geologie, Mineralogie und Paläontologie*. Berlin, Leipzig: W. De Gruyter & Co., 1928.

Shemella, P. W. *Dictionary of Mineralogical Names and Associated Terminology, 1975*. Friends of Mineralogy in Region 3, 1975.

Strunz, Hugo. *Mineralogische Tabellen*, Fifth Ed. Leipzig: Akademische Verlagsgesellschaft, 1970.

Ure, Andrew. *A Dictionary of Chemistry and Mineralogy, with Their Applications*, Fourth Ed. London: T. Tegg & Son, 1835.

von Kobell, Wolfgang Franz. *Die Mineral-Namen und die Mineralogische Nomenklatur*. Munich: J. G. Cotta, 1853.

Whewell, William. *An Essay on Mineralogical Classification and Nomenclature*. Cambridge: 1828.

Zappe, Joseph R. *Mineralogisches Handlexicon*, 3 Volumes. Wien: 1817.

Index

This index only includes *Part I—Mineral Names: A Discussion.*